Raw and Finished Materials

A Concise Guide to Properties and Applications

RAW AND FINISHED MATERIALS

A Concise Guide to Properties and Applications

Brian Dereu

MOMENTUM PRESS

CONTENTS

INTRODUCTION

A basic understanding and knowledge of materials is rudimentary to both the accomplished tradesman as well as the draftsman. The subject of materials, however, is not generally taught in a vocational setting. It will be shown that a length of brass is not just a length of brass, but an example of a specific alloy that determines whether it can be more easily machined, or rather formed. Yet, for each of the basic materials listed, entire texts have been written and published. It is not, therefore, my intention to delve deeply into such matters as failure analysis or to dissect various exotic composites. Rather, this is a concise text that will introduce the most common materials along with their properties and uses, which both the student and journeyman will find useful. Even as much, there are almost 500 alloys of wrought aluminum registered to the Aluminum Association of the United States, though I have selected only seven to list in this text. No assumption has been made concerning the previous education of the reader, and everything has been written from the ground up, without jargon.

The text has four main parts;

Part I   Metals
Part II   Plastics
Part III   Woods
Part IV   Ceramics, Composites, and Miscellaneous Materials

The format has been written to move from the broad to the narrow. Taking "Metals" as an example, it is broken down into:

- Ferrous Metals
- Non-Ferrous Metals

Continuing on with "Nonferrous Metals," we have:

- Aluminum and its Alloys
- Copper and its Alloys
- Nickel and its Alloys
- Miscellaneous Nonferrous metals

Under "Copper and its Alloys," we will find:

- Straight Coppers
- Brasses
- Bronzes

This is where "Alloy C36000, Free Cutting Brass" would be found. In this entry, C36000 is discussed and listed as to its elemental makeup, properties, uses, and special considerations.

For rapid reference, a comprehensive index, including common and trademarked names, is included. For example, Lexan® and polycarbonate will reference the same entry. By the same token, many metals have common trade names as well as UNS numbers, and both are indexed, as in O-1 tool steel and UNS T31501.

A variety of charts and graphs are found throughout that compile and compare information within visual packets that help visualize relationships between or within materials or processes. Photographs are also used as complementary visual aids. It is my intention and hope that this book will help fill the gap in general materials knowledge and be a useful tool, not only for teaching the tradesman in training but also as a quick reference for the accomplished journeyman.

Brian Dereu

Metals

1

Ferrous Metals

1.1 IRON

1.1.1 Gray Cast Iron

Gray cast iron, or simply *gray iron,* is a common engineering material consisting of iron with a typical carbon content of 2.5 to 3.5 percent and a silicon content of around 2 percent. Other elements, including manganese and copper, may be introduced to give specific properties or found as impurities (in the case of sulfur). Gray iron can be poured at the lowest temperature of any ferrous material and at the same time will have the lowest viscosity, enabling it to fill tiny voids and intricate shapes. The carbon in gray iron is mostly in the form of graphite, which gives it excellent machining properties. It can be machined dry, but coolant may be used to keep the dust and small chips down. The graphite also gives it superb anti-seizing qualities, giving rise to its use as gasoline engine block cylinder walls and machine tool travel ways. Gray iron has fairly low tensile strength and ductility and cannot withstand shock to any degree without cracking. Despite its high concentration of flake graphite, it is successfully welded and brazed.

**Table 1.1 Gray Cast Iron Properties (Annealed, 1-in°
Round Stock)**[*]

	Metric	English
Hardness, Rockwell B	97	97
Tensile Strength, Ultimate	$\geq$276 MPa	$\geq$40000 psi
Machinability (based on AISI 1212 as 100%)	70–90%	70–90%

Chemical Composition

	Metric	English
Carbon	3.25–3.50%	3.25–3.50%
Chromium	0.050–0.45%	0.050–0.45%
Manganese	0.50–0.90%	0.50–0.90%
Molybdenum	0.050–0.10%	0.050–0.10%
Phosphorous	$\leq$0.12%	$\leq$0.12%
Silicon	1.80–2.30%	1.80–2.30%
Sulfur	$\leq$0.15%	$\leq$0.15%

[*]Selected data from the MatWeb search database reprinted with permission from MatWeb, www.matweb.com.

1.1.2 Ductile Cast Iron (Nodular Iron)

Ductile iron is nearly similar in chemical makeup to gray iron
but has drastically different mechanical and physical properties
due to the addition of magnesium. The added magnesium (or
sometimes cerium) forms the carbon into spherical nodules,
unlike the flakes found in gray iron. The carbon nodules act to
stop any forming cracks, giving the iron its ductility. In fact,
some ductile irons have elongation yields in excess of 25 per-

cent before failure, whereas gray iron will fail without any appreciative stretch. Ductile iron is machinable, but some grades are harder than others, and carbide tooling may be required.

By far most of the ductile iron used is in buried pipe for water and sewer use. Iron pipe competes with steel, concrete, and plastic pipe for these purposes, and each has advantages and disadvantages based on variables such as diameter, length of run, and so forth. Other uses of ductile iron include automotive manifolds, brake drums, and external accessories.

Table 1.2   Ductile Cast Iron Properties (Annealed, 1-in Round Stock)[*]

	Metric	English
Tensile Strength, Ultimate	≥414 MPa	≥60,000 psi
Tensile Strength, Yield	≥276 MPa	≥40,000 psi
Elongation at Break	≥18.0%	≥18.0%
Chemical Composition		
Carbon	3.50–3.90%	3.50–3.90%
Chromium	≤0.050%	≤0.050%
Iron	95.0%	95.0%
Manganese	0.15–0.35%	0.15–0.35%
Phosphorous	≤0.050%	≤0.050%
Silicon	2.25–2.75%	2.25–2.75%
Sulfur	0.010–0.025%	0.010–0.025%

[*]Selected data from the MatWeb search database reprinted with permission from MatWeb, www.matweb.com.

1.1.3 White Cast Iron

White cast iron, or simply *white iron,* is named from its color appearance when freshly fractured. When white iron is poured and solidifies from its liquid form, the carbon content does not form graphite as in gray iron but stays in solution with the iron creating iron carbides. These carbides result in an extremely hard, brittle iron that is practically impossible to conventionally machine and can reach a hardness of RC 62+. White iron is generally used "as cast," but it can be ground with abrasive wheels. Welding this material is also very difficult and not carried out except in repair work. Although a very brittle material, white iron is extremely hard and has good compression strength. It finds uses in components for rock and coal crushers, grinding mills, and abrasive material pump impellers.

Table 1.3   White Cast Iron Properties (Annealed, 1-in Round Stock)[*]

	Metric	English
Hardness, Rockwell C	47.1–54.4	47.1–54.4
Tensile Strength, Ultimate	345–689 MPa	50,000–100,000 psi
Modulus of Elasticity	165–179 GPa	24,000–26,000 ksi
Chemical Composition		
Carbon	2.00–3.70%	2.00–3.70%
Chromium	1.10–28.0%	1.10–28.0%
Iron	65.0–88.0%	65.0–88.0%
Manganese	0.500–2.00%	0.500–2.00%
Molybdenum	0.500–3.50%	0.500–3.50%

Table 1.3 White Cast Iron Properties (Annealed, 1-in Round Stock)[*] *(cont.)*

	Metric	English
Phosphorous	0.100–0.300%	0.100–0.300%
Silicon	0.800–2.20%	0.800–2.20%
Sulfur	0.0600–0.150%	0.0600–0.150%

[*]Selected data from the MatWeb search database reprinted with permission from MatWeb, www.matweb.com.

1.2 STEELS

1.2.1 Steel

Steel is an alloy of iron combined with other elements. The most common steels contain iron and carbon in varying amounts, giving different properties. Carbon steel is often called *straight steel,* as opposed to alloy steels that contain carbon along with other elements. There are literally hundreds of steel alloys in use today, varying from relatively soft low carbon automobile panels to rock-hard high speed steel cutting tools. In between these two extremes are bridges, ships, rail track, and countless other items made of various steels. Carbon steel itself has varying properties of hardenability, ductility, and malleability depending on the amount of carbon used. Straight high carbon steel is unmatched in hardenability of all steels, including alloy steels but, when in its super hard state, it is brittle and unyielding. The addition of various alloying elements gives good hardness and at the same time provides strength, resistance to wear, toughness, and ductility.

Straight carbon steel can be divided into three categories, depending upon the amount of carbon added: low carbon, medium carbon, and high carbon.

1.2.1.1 Low Carbon Steel

Steel containing less than 0.3 percent carbon is known as low carbon or *mild* steel. The low amount of carbon makes this steel non-hardenable, and therefore it is used in its soft state. Mild steel is the most malleable of the steels and is often rolled into sheet metal for uses including automobile panels, appliance shells, radio chassis, and street signs. When wrought into round stock, low carbon steel is used to make bolts, nuts, and other hardware. At the upper end are structural steels including A36, a common mild steel containing 0.26 percent carbon. Mild steel is also easily welded.

1.2.1.2 Medium Carbon Steel

Steel containing between 0.3 and 0.8 percent carbon is known as *medium carbon steel*. With this amount of carbon, steel becomes hardenable through heat treating, but medium carbon steel possesses good strength and toughness even in its soft state, in which it is often used. As its hardness and strength increase, its malleability and ductility decrease, limiting its use for cold working. Uses include gears, crankshafts, and other machine parts needing good strength and machinability.

1.2.1.3 High Carbon Steel

As the carbon content in steel increases, maximum hardness potential also increases. High carbon steel contains over 0.8 percent carbon and is used for cutting tools and blades where rock-hard, keen cutting edges are needed. To hold an edge and provide against wear, a very high degree of hardness is needed. Other items (e.g., punches and chisels) require a combination of hardness and resiliency to do their work without chipping or shattering. Once heat treated, these steels can be annealed to varying degrees of hardness for their specific purposes.

1.2.1.4 Alloy Steels

Although carbon steel makes up the bulk of all steels used, there are hundreds of other alloys used to lesser degrees. These specialty steels have properties above and beyond carbon steels and include the tool steels and high-speed steels. There are approximately 12 metallic and nonmetallic elements that are added to iron to make alloy steels in different combinations and amounts to give the alloys their specific properties. These properties include hardness, toughness, abrasion and corrosion resistance, and the ability to hold those properties at high heat (red hardness).

1.2.2 Steel Designations in the AISI/SAE System

The AISI/SAE steel numbering system uses a (mainly) four-digit number to classify steels based on their chemical composition. The first and second digits typify the major series as described below. The second two digits refer to the carbon content of the steel, expressed in hundredths of one percent. Although the AISI/SAE system was integrated into the UNS system of numbering, the AISI/SAE system is still favored due to its familiarity of use.

Carbon Steels

- 10xx Plain carbon steel
- 11xx Resulfurized steel
- 12xx Resulfurized and rephosphorized steel
- 15xx Plain carbon with elevated manganese levels from 1.00 to 1.65 percent

Manganese Steels

- 13xx

Nickel Steels

- 23xx
- 25xx

Nickel-Chrome Steels

- 31xx
- 32xx
- 33xx
- 34xx

Molybdenum Steels

- 40xx
- 44xx

Chromium-Molybdenum Steels

- 41xx

Nickel-Chromium-Molybdenum Steels

- 43xx
- 47xx
- 81xx
- 86xx
- 87xx
- 88xx
- 93xx
- 94xx

- 97xx
- 98xx

Nickel-Molybdenum Steels

- 46xx
- 48xx

Chromium Steels

- 50xx
- 51xx
- 52xx

Chromium-Vanadium Steels

- 61xx

Tungsten-Chromium Steels

- 72xx

Silicon-Manganese Steels

- 92xx

Leaded Steels

- xxLxx

Boron Steels

- xxBxx

1.2.3 The Unified Numbering System (UNS)

The UNS has been created by a joint effort of the Society of Automotive Engineers (SAE) and the American Society for Testing and Materials (ASTM) to associate the various alloy designation systems that are in use by industry. Before the utilization of the UNS, each major trade association (including the American Aluminum Association, the Copper Development Association, and the American Iron and Steel Institute) carried their own designations. These individual systems, though still widely used, present problems with overlapping numbers and miscommunication involving trade names.

Ultimately, the need for a centralized system of alloy identification and callout was needed, and in 1974 The UNS was established. The standards set by the UNS are universal, encompassing all metals and alloys, and are expandable as new alloys are produced. Having a single system available not only prevents errors in communication, it also forms a central database for rapid reference of any alloy. In forming the UNS system, older alloy designations were absorbed into its format when possible; i.e., American Aluminum alloy "6061" is designated as "A96061" under the UNS system. As shown in the example "A96061," the UNS designation consists of a letter prefix, which is often descriptive of the major metal (A for aluminum), followed by a five-digit number. In aluminum UNS numbering, the "9" designates a wrought alloy.

It is to be noted that UNS designations only specify the chemical makeup of an alloy, not its temper or hardness.

The following prefixes of the UNS system are used to designate the metal family:

- Axxxxx   Aluminum and its alloys
- Cxxxxx   Copper and its alloys
- Exxxxx   Rare earth and similar materials
- Fxxxxx   Cast irons
- Gxxxxx   AISI and SAE carbon and alloy steels
- Hxxxxx   AISI and SAE "H" steels
- Jxxxxx   Cast steels excluding tool steels
- Kxxxxx   Miscellaneous steels and other ferrous alloys

- Lxxxxx   Low melting metals
- Mxxxxx   Miscellaneous nonferrous metals
- Nxxxxx   Nickel and its alloys
- Pxxxxx   Precious metals
- Rxxxxx   Reactive and refractory metals
- Sxxxxx   Heat and corrosion resistant steels, including stainless steels and iron based superalloys
- Txxxxx   Tool steels
- Wxxxxx   Welding filler metals
- Zxxxxx   Zinc and its alloys

In practice, many of the older systems are still in use, including steel grades in the AISI/SAE structure, American Aluminum's International Alloy Designation System (IADS), and the three-digit system from the Copper Development Association for copper and its alloys (see Fig. 1.1).

1.2.3.1 Low Carbon Steel SAE 1018 (UNS G10180)

SAE 1018 low carbon steel is one of the more common steel grades and is used when high strength is not an issue. SAE 1018

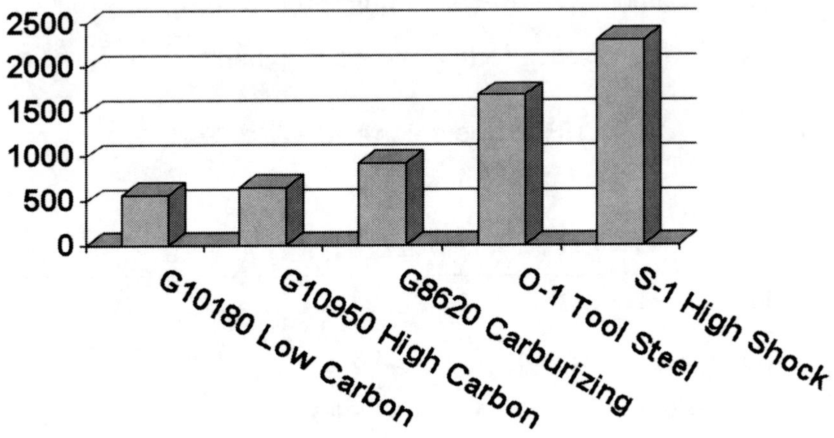

FIGURE 1.1 Ultimate tensile strengths, various steels (MPa).

can be easily welded and cold formed without difficulty, and machinability is rated at 65 percent of SAE 1212. It is not heat treatable by ordinary methods, due to its low carbon content, but it can be case hardened successfully. SAE 1018 steel is available in all standard shapes and gauges. When machining 1018, it is difficult to obtain a good finish, as the work tends to gall and build up on the cutting tool. It is, however, inexpensive and readily available from all supply houses, making it a popular choice for general work. For extensive machining operations, a better choice might be SAE 1212 or SAE 1215. These two grades will machine much more easily with less pressure required, producing broken, noncontinuous chips (known as "free cutting") and leaving better finishes. The ultimate steel for machining, however, is 12Lxx. The "L" designates added lead, which acts similarly in steel as in brass to lubricate the cutting tool/work boundary. Alloy 12L14 has a machinability rating of 160 percent of SAE 1212, requires the least pressure to machine of all steels, and leaves the best finishes. It has two drawbacks, however: its high susceptibility to corrosion and its lead content.

Table 1.4   SAE 1018 Properties (Annealed, 1-in Round Stock)[*]

	Metric	English
Hardness, Rockwell B	84	84
Tensile Strength, Ultimate	565 MPa	81,900 psi
Tensile Strength, Yield	485 MPa	70,300 psi
Elongation at Break	20.0%	20.0%

Table 1.4   SAE 1018 Properties (Annealed, 1-in Round Stock)[*] *(cont.)*

	Metric	English
Modulus of Elasticity	200 GPa	29,000 ksi
Machinability	65%	65%
Thermal Conductivity	51.9 W/m·K	360 Btu-in/hr·ft^2·°F

Chemical Composition

	Metric	English
Carbon	0.14–0.20%	0.14–0.20%
Iron	98.81–99.26%	98.81–99.26%
Manganese	0.60–0.90%	0.60–0.90%
Phosphorous	≤0.040%	≤0.040%
Sulfur	≤0.050%	≤0.050%

[*]Selected data from the MatWeb search database reprinted with permission from MatWeb, www.matweb.com.

1.2.3.2 *UNS G12144 Low Carbon Leaded Free Machining Steel (AISI 12L14)*

12L14 is the easiest machining steel, having a rating of 160 percent over AISI 1212. Its lead content varies among manufacturers and can fluctuate between 0.15 and 0.35 percent. Two large consumers of leaded steels are screw machine users and hobbyists. Whether using manual Swiss-type screw machines or modern CNC lathes, high-production shops need every angle to remain competitive. Using the best machining steel will shave time off each part produced, meaning more parts produced per time unit. 12L14 takes less horsepower and allows deeper cuts at faster rates than any other steel, all the while leaving a very descent finish. These same properties appeal to the hobby machinist who tends to have smaller, less rigid machinery that will still cut 12L14 with no problems.

Screw machine products made from 12L14 include pins, shafts, bolts, studs, and similar hardware. 12L14 steel is not just confined to round stock, of course, and square and flat stock is often used in the milling machine to produce precision fixturing, jigs, and tools. It can be either plated or case hardened with ease, but is too low in carbon for solution heat treating. One downfall of unplated 12L14 is its lack of corrosion resistance. Most items produced from this material are plated in nickel, chrome, or brass for both corrosion resistance and eye appeal.

Table 1.5   UNS G12144 Properties (Annealed, 1-in Round Stock)[*]

	Metric	English
Hardness, Rockwell B	84	84
Tensile Strength, Ultimate	540 MPa	78,300 psi
Tensile Strength, Yield	415 MPa	60,200 psi
Elongation at Break	10.0%	10.0%
Modulus of Elasticity	200 GPa	29000 ksi
Machinability (based on AISI 1212 as 100%)	160%	160%
Thermal Conductivity	51.9 W/m·K	360 Btu-in/hr·ft^2·°F

Chemical Composition

	Metric	English
Carbon	≤0.15%	≤0.15%

Table 1.5   UNS G12144 Properties (Annealed, 1-in Round Stock)[*] *(cont.)*

	Metric	English
Iron	97.91–98.7%	97.91–98.7%
Manganese	0.85–1.15%	0.85–1.15%
Phosphorous	0.040–0.090%	0.040–0.090%
Sulfur	0.260–0.35%	0.260–0.35%

[*]Selected data from the MatWeb search database reprinted with permission from MatWeb, www.matweb.com.

1.2.3.3 *AISI 1050 Medium Carbon Steel (UNS G10500)*

AISA 1050 medium carbon steel, while still having a high enough carbon content to be heat treatable, it does not have the hardness potential of a higher carbon steel such as 1095. Its lower carbon content gives it better toughness and ductility, and is a good choice when a lower hardness will suffice, having better machinability and weldability as well.

Table 1.6   *AISI 1050* Properties (Annealed, 1-in Round Stock)[*]

	Metric	English
Hardness, Rockwell B	90	90
Tensile Strength, Ultimate	655 MPa	95,000 psi
Tensile Strength, Yield	550 MPa	79,800 psi
Elongation at Break	10.0%	10.0%
Modulus of Elasticity	205 GPa	29,700 ksi
Machinability (based on AISI 1212 as 100%)	55%	55%

Table 1.6   *AISI 1050* Properties (Annealed, 1-in Round Stock)[*] *(cont.)*

	Metric	English
Thermal Conductivity	49.8 W/m·K	346 Btu-in/hr·ft^2·°F

Chemical Composition		
Carbon	0.470–0.55%	0.470–0.55%
Iron	98.46–98.92%	98.46–98.92%
Manganese	0.60–0.90%	0.60–0.90%
Phosphorous	≤0.040%	≤0.040%
Sulfur	≤0.050%	≤0.050%

[*]Selected data from the MatWeb search database reprinted with permission from MatWeb, www.matweb.com.

1.2.3.4 High Carbon Steel SAE 1095 (UNS G10950)

1095 high carbon steel is a favorite material for knife blades due to its high hardenability, strength, and ability to hold an edge. It is also a common spring steel and is available either tempered (hardened) or in an annealed (soft) state for making flat or leaf springs. When sold as hardened, drawn wire, it is knows as "piano" or "music" wire, and is used to wind coil springs and in musical instruments. Being straight steel, it corrodes easily and must be safeguarded. 1095 is difficult to machine, even in its annealed state. When welding 1095 steel, it must be annealed afterward to counter the hardening from the high welding heat. Otherwise, a brittle weld area will be left that can result in failure.

Table 1.7 *SAE 1095* **Properties (Annealed, 1-in Round Stock)**[*]

	Metric	English
Hardness, Rockwell B 91	91	91
Tensile Strength, Ultimate	655 MPa	95000 psi
Tensile Strength, Yield	380 MPa	55,100 psi
Elongation at Break	13.0%	13.0%
Modulus of Elasticity	200 GPa	29000 ksi
Thermal Conductivity	51.9 W/m·K	360 Btu-in/hr·ft^2·°F

Chemical Composition

	Metric	English
Carbon	0.90–1.03%	0.90–1.03%
Iron	98.38–98.8%	98.38–98.8%
Manganese	0.30–0.50%	0.30–0.50%
Phosphorous	≤0.040%	≤0.040%
Sulfur	≤0.050%	≤0.050%

[*]Selected data from the MatWeb search database reprinted with permission from MatWeb, www.matweb.com.

1.2.3.5 A36 Structural Steel (UNS K02600)

A36 steel is the most common structural steel used in the U.S.A. It is available in plate, sheet, and structural shapes and is easily welded by all common methods. It is a fairly ductile steel and will "give" a bit in high winds when used in tall structures, while a more brittle steel used in similar applications could fail through breakage. A36 is difficult to machine and relatively soft and gummy when compared to other steels. When holes are called for in structural shapes, they are punched ahead of time at the mill. In the machine shop, carbide tooling can successfully machine A36, but it is difficult at best. Not only is it gummy inside, but its hot rolled scaly finish is abrasive on tools.

Table 1.8   A36 Properties (Annealed, 1-in Round Stock)[*]

	Metric	English
Tensile Strength, Ultimate	400–550 MPa	58,000–79,800 psi
Tensile Strength, Yield	≥250 MPa	≥36,300 psi
Elongation at Break	≥23.0%	≥23.0%
Modulus of Elasticity	200 GPa	29,000 ksi
Thermal Conductivity	51.9 W/m·K	360 Btu-in/hr·ft^2·°F
Chemical Composition		
Carbon	0.260%	0.260%
Iron	99.0%	99.0%
Manganese	0.75%	0.75%
Phosphorous	≤0.040%	≤0.040%
Sulfur	≤0.050%	≤0.050%

[*]Selected data from the MatWeb search database reprinted with permission from MatWeb, www.matweb.com.

1.2.3.6 *ASTM A366 Cold Rolled Sheet Steel (UNS G10080) and ASTM A653 Galvanized Sheet Steel*

A366 cold rolled sheet steel is used for appliance shells, radio chassis, and other stamped parts. It is the common steel used as "sheet metal" and is available in all thickness gauges. The cold rolled finish on A366 is very smooth and scale free, unlike hot rolled sheet, and it is available from the mill in plain (matte) to luster finishes. ASTM A653 steel is similar in makeup but is hot dipped galvanized for corrosion protection. The visible crystals, or *spangle,* on a sheet of galvanized steel vary in size according to the amount of lead (as an impurity) in the bath and production methods. A finer, or totally absent, visible spangle will allow the best finishes when the metal is painted and may be called for by the fabricator. The spangle-free finish is not as durable, however, when extensive drawing or other cold working operations are to be done.

Table 1.9   A366 Cold Rolled Steel Properties (Annealed, 1-in Round Stock)[*]

	Metric	English
Hardness, Rockwell B	55	55
Tensile Strength, Ultimate	340 MPa	49,300 psi
Tensile Strength, Yield	285 MPa	41,300 psi
Elongation at Break	20.0%	20.0%
Modulus of Elasticity	200 GPa	29,000 ksi
Machinability	55%	55%
Thermal Conductivity	65.2 W/m·K	452 Btu-in/hr·ft^2·°F

Chemical Composition

	Metric	English
Carbon	≤0.15%	≤0.15%
Iron	99.0%	99.0%
Manganese	≤0.60%	≤0.60%
Phosphorous	≤0.035%	≤0.035%
Sulfur	≤0.040%	≤0.040%

[*]Selected data from the MatWeb search database reprinted with permission from MatWeb, www.matweb.com.

1.2.3.7 4140 Chrome Molybdenum Steel (UNS G41400)

More rifle, shotgun, and pistol barrels are made from 4140 chrome moly steel than any other metal, due to its various properties. Firearm barrels must have a combination of strength and toughness to contain the tens of thousands of pounds pressure in the chamber when each shot is fired. The steel also must be ductile and resilient to prevent cracking and splitting. 4140 has all of these properties. When barrels are made, the steel is preheat treated to around RC 25 to 35, depending on the manufacturer and barrel type. At this hardness, 4140 is still easily machinable and can be drilled, rifled, chambered, and threaded. A final heat normalizing relieves the internal stresses that drilling and rifling left behind, and the barrel is now ready to be polished and blued.

Other uses of 4140 include general machine parts, including gears, shafting, and axles. Its 1 percent chromium content slightly improves its resistance to corrosion over straight carbon steel.

Table 1.10   4140 Chrome Molybdenum Steel Properties (Annealed, 1-in Round Stock)[*]

	Metric	English
Hardness, Rockwell B	98	98
Tensile Strength, Ultimate	324 MPa	47,000 psi
Modulus of Elasticity	330 GPa	47,900 ksi
Thermal Conductivity	138 W/m·K	958 Btu-in/hr·ft^2·°F
Chemical Composition		
Carbon	0.38–0.43%	0.38–0.43%
Chromium	0.80–1.10%	0.80–1.10%
Iron	96.785–97.77%	96.785–97.77%

Table 1.10   4140 Chrome Molybdenum Steel Properties (Annealed, 1-in Round Stock)[*] (cont.)

	Metric	English
Manganese	0.75–1.0%	0.75–1.0%
Molybdenum	0.15–0.25%	0.15–0.25%
Phosphorous	≤0.035%	≤0.035%
Silicon	0.15–0.30%	0.15–0.30%
Sulfur	≤0.040%	≤0.040%

[*]Selected data from the MatWeb search database reprinted with permission from MatWeb, www.matweb.com.

1.2.3.8 AISI 8620 Carburizing Steel (UNS G86200)

8620 is a nickel, chrome and molybdenum steel that is especially susceptible to carburizing (case hardening), and can produce a skin in excess of RC60. Also unlike case hardened low carbon steel, 8620 will have an interior hardness of RC 30 or so after carburizing. It is a favorite material for gears that need good strength and a high exterior hardness for metal-to-metal contact. Other uses include splined shafts, cam shafts, cam lobes and other toothed or high wear parts. 8620 steel was also used for the bolt and receiver on the M-1 Garand U.S. infantry rifle.

Table 1.11   8620 Carburizing Steel Properties (Annealed, 1-in Round Stock)[*]

	Metric	English
Hardness, Rockwell B	99	99
Tensile Strength, Ultimate	931 MPa	135,000 psi
Tensile Strength, Yield	758 MPa	110,000 psi

Table 1.11   8620 Carburizing Steel Properties (Annealed, 1-in Round Stock)[*] **(cont.)**

	Metric	English
Elongation at Break	16.0%	16.0%
Modulus of Elasticity	205 GPa	29,700 ksi
Machinability (based on AISI 1212 as 100%)	65%	65%
Thermal Conductivity	46.6 W/m·K	323 Btu-in/hr·ft^2·°F

Chemical Composition

	Metric	English
Carbon	0.180–0.230%	0.180–0.230%
Chromium	0.400–0.600%	0.400–0.600%
Iron	96.895–98.02%	96.895–98.02%
Manganese	0.700–0.900%	0.700–0.900%
Molybdenum	0.150–0.250%	0.150–0.250%
Phosphorous	≤0.0350%	≤0.0350%
Silicon	0.150–0.350%	0.150–0.350%
Sulfur	≤0.0400%	≤0.0400%

[*]Selected data from the MatWeb search database reprinted with permission from MatWeb, www.matweb.com.

1.2.3.9 M2 High Speed Steel (UNS T11302)

M2 is a general-purpose high speed steel (HSS) and is a common material for twist drills, end mills, reamers, and lathe tools (Fig. 1.2). It provides good value for minimum cost and, although M2 is at the bottom of the HSS family, it out-performs carbon steel tools by far. More important than the chemistry of the HSS is the proper grinding of the tool with correct geometry and quality in manufacturing. M2 is suited for mild steel, the easier cutting nonferrous materials, nonabrasive plastics, and wood. For high carbon and tool steels, an M42 Cobalt HSS has better high heat resistance, durability and hardness.

Table 1.12   M2 High Speed Steel Properties (Annealed, 1-in Round Stock)

	Metric	English
Hardness, Rockwell C	64.8	64.8
Modulus of Elasticity	207 GPa	30,000 ksi
Machinability (based on AISI 1212 as 100%)	45.0–50.0%	45.0–50.0%

Chemical Composition

	Metric	English
Carbon	0.840%	0.840%
Chromium	4.15%	4.15%
Iron	83%	83%
Molybdenum	5.0%	5.0%
Silicon	0.30%	0.30%
Tungsten	6.65%	6.65%
Vanadium	1.85%	1.85%

1.2.3.10 M42 Cobalt HSS (UNS T11342)

M42 high speed steel (HSS) is used in the production of drills, end mills, reamers, and lathe tools. Its high cobalt content (~8 percent) gives it special properties above regular M2 HSS at only a slightly higher cost. M42 is used for cutting high carbon steels, tool steels, titanium, and other difficult-to-machine materials. In cutting softer materials like wood, it outlasts M2 HSS and is a favorite material for planer blades.

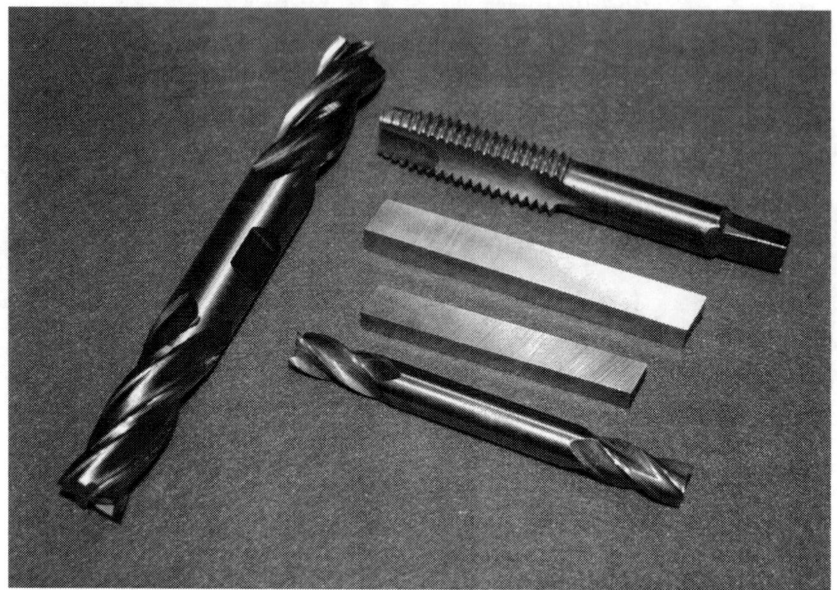

FIGURE 1.2 An assortment of high speed steel (HSS) cutting tools.

Table 1.13   M42 Cobalt HSS Properties (Annealed, 1-in Round Stock)[*]

	Metric	English
Hardness, Rockwell C	64.0–66.0	64.0–66.0
Modulus of Elasticity	206.8 GPa	29990 ksi

Chemical Composition

	Metric	English
Carbon	1.10%	1.10%
Chromium	3.75%	3.75%
Cobalt	8.25%	8.25%
Iron	74.69%	74.69%
Molybdenum	9.50%	9.50%
Sulfur	0.060%	0.060%

Table 1.13   M42 Cobalt HSS Properties (Annealed, 1-in Round Stock)[*] *(cont.)*

	Metric	English
Tungsten	1.50%	1.50%
Vanadium	1.15%	1.15%

[*]Selected data from the MatWeb search database reprinted with permission from MatWeb, www.matweb.com.

1.2.3.11 AISI T1 Tungsten High Speed Steel (UNS T12001)

T1 high speed steel (HSS) is high-tungsten tool steel having excellent wear and toughness properties. It is an older, nearly obsolete steel and has largely been replaced by M2 HSS. It still is favored over M2, however, for planer and jointer blades used on hardwoods in both large mills and home wood shops. T1 hardens at a higher temperature than M2, meaning that it can withstand higher operating temperatures without failing. Its higher tungsten content allows easier grinding, also.

Table 1.14   AISI T1 Properties (Annealed, 1-in Round Stock)[*]

	Metric	English
Hardness, Rockwell C	63.0–65.0	63.0–65.0
Thermal Conductivity	20.0 W/m·K	139 Btu-in/hr·ft^2·°F

Chemical Composition

	Metric	English
Carbon	0.750%	0.750%
Chromium	4.13%	4.13%
Iron	76.02%	76.02%

Table 1.14   AISI T1 Properties (Annealed, 1-in Round Stock)[*] (cont.)

	Metric	English
Tungsten	18.0%	18.0%
Vanadium	1.10%	1.10%

[*]Selected data from the MatWeb search database reprinted with permission from MatWeb, www.matweb.com.

1.2.3.12 O-1 Tool Steel (UNS T31501)

O-1 is an oil hardening, general-purposed tool steel available from all supply houses in round shapes as well as flat sections. O-1 steel can be heat treated at relatively low temperatures, as compared to air hardening steels, and is popular in the machine shop for making small "as-needed" tools, including drills, punches, dies, and so forth. O-1 steel round stock is commonly known as "drill rod," but this steel lacks the durability and edge-retaining qualities of high speed steels, which are a much better choice for twist drills. Although the manufacturers of these steels publish very stringent heat-treating and annealing temperature charts, a trained individual with a can of quenching oil, an oxy-acetylene torch, and a good eye for color can successfully heat treat and temper small parts and tools.

Table 1.15   O-1 Tool Steel Properties (Annealed, 1-in Round Stock)[*]

	Metric	English
Hardness, Rockwell B	50	50
Tensile Strength, Ultimate	1690 MPa	245,000 psi
Tensile Strength, Yield	1500 MPa	218,000 psi

Table 1.15   O-1 Tool Steel Properties (Annealed, 1-in Round Stock)[*] *(cont.)*

	Metric	English
Elongation at Break	0.000%	0.000%
Modulus of Elasticity	214 GPa	31,000 ksi
Machinability (based on AISI 1212 as 100%)	40%	40%
Thermal Conductivity	32.0 W/m·K	222 Btu-in/hr·ft^2·°F

Chemical Composition

	Metric	English
Carbon	0.85–1.0%	0.85–1.0%
Chromium	0.50%	0.50%
Iron	96.0%	96.0%
Manganese	1.20%	1.20%
Phosphorous	≤0.030%	≤0.030%
Silicon	≤0.50%	≤0.50%
Sulfur	≤0.030%	≤0.030%

[*]Selected data from the MatWeb search database reprinted with permission from MatWeb, www.matweb.com.

1.2.3.13 D2 Tool Steel (T30402)

D2 is a high carbon, high chromium, air hardening tool steel with excellent abrasion and wear resistance. Its high chromium content (11–13 percent) gives it good corrosion resistance relative to other steels, but it will still corrode and rust if not protected. D2 is a favorite steel of custom and commercial knife makers. Its abrasion resistance enables it to hold an edge better than other knife steels, particularly 01 and 440C. D2 is very difficult to machine and properly heat treat, however, as compared to 01, so properties are mixed.

Besides the use in cutlery blades, D2 steel is used for many industrial tools, including coining dies, extrusion dies, forming rolls, and slitting cutters. Its good wear resistance gives tools long life, which can more than makeup for its poor machining characteristics.

Table 1.16   D2 Tool Steel Properties (Annealed, 1-in Round Stock)[*]

	Metric	English
Hardness, Rockwell C	62	62
Chemical Composition		
Carbon	1.50%	1.50%
Chromium	12.0%	12.0%
Iron	85.0%	85.0%
Molybdenum	1.0%	1.0%
Vanadium	0.80%	0.80%

[*]Selected data from the MatWeb search database reprinted with permission from MatWeb, www.matweb.com.

1.2.3.14 P20 Tool Steel (UNS T51620)

P20 tool steel is a nickel, chrome, molybdenum mold steel that is supplied by the manufacturer in a hardened state (usually near RC 30). In this condition, it is still machinable and usually requires no subsequent hardening. P20 is a very "clean" steel, allowing it to be polished to a very high finish. It is generally used for plastic injection molds and die cast molds for zinc, tin, and similar metals. It has good high heat hardness and wear resistance.

Table 1.17   P20 Tool Steel Properties (Annealed, 1-in Round Stock)[*]

	Metric	English
Hardness, Rockwell C	30	30
Tensile Strength, Ultimate	965–1030 MPa	140,000–150,000 psi
Tensile Strength, Yield	827–862 MPa	120,000–125,000 psi
Elongation at Break	20.0%	20.0%
Modulus of Elasticity	205 GPa	29,700 ksi
Thermal Conductivity	29.0–34.0 W/m·K	201–236 Btu-in/hr·ft^2·°F

Chemical Composition

	Metric	English
Carbon	0.280–0.40%	0.280–0.40%
Chromium	1.40–2.0%	1.40–2.0%
Iron	97.0%	97.0%
Manganese	0.60–1.0%	0.60–1.0%
Molybdenum	0.30–0.55%	0.30–0.55%
Phosphorous	≤0.030%	≤0.030%
Silicon	0.20–0.80%	0.20–0.80%
Sulfur	≤0.030%	≤0.030%

[*]Selected data from the MatWeb search database reprinted with permission from MatWeb, www.matweb.com.

1.2.3.15 T41901 S1 High Shock Tool Steel (UNS T41901)

S1 is a medium carbon, high chromium and tungsten, oil hardening tool steel exhibiting a high degree of shock resistance and good strength and hardenability (to over RC60). As with all tool steels, it should be preheated before welding to prevent cracking and distorting. The machinability of S1 is fair to good, rating 70% of AISI 1212. It is used mainly for its high shock resistance in chisels, hand punches, rivet sets, and pneumatic tools. Other uses include shear blades, coining dies, and gear cutting hobs. When high shock is not an issue, other tool steels may be better choices due to S1's poor abrasive and wear resistance.

Table 1.18   S1 High Shock Tool Steel Properties (Annealed, 1-in Round Stock)[*]

	Metric	English
Hardness, Rockwell C	58	58
Tensile Strength, Ultimate	2310 MPa	335,000 psi
Tensile Strength, Yield	1750 MPa	254,000 psi
Elongation at Break	9.00%	9.00%

Chemical Composition

	Metric	English
Carbon	0.40–0.55%	0.40–0.55%
Chromium	1.40%	1.40%
Iron	94.0%	94.0%
Manganese	0.25%	0.25%
Molybdenum	≤0.50%	≤0.50%
Phosphorous	≤0.030%	≤0.030%
Silicon	0.680%	0.680%
Sulfur	≤0.030%	≤0.030%
Tungsten	2.25%	2.25%
Vanadium	0.23%	0.23%

[*]Selected data from the MatWeb search database reprinted with permission from MatWeb, www.matweb.com.

1.2.4 **Stainless Steels**

Stainless steels are alloys that possess a great degree of corrosion resistance due to their high chromium content (usually over 10 percent). The chromium reacts with the oxygen to form chromium oxide; a tough protective layer that prevents corrosion underneath. Chromium oxide is self-healing and will quickly form on the exterior if the metal is scratched or abraded off. Stainless steels are tough, gummy, and generally difficult to machine. There are three main grades of stainless steel: austenitic, martensitic, and ferritic. In each grade, there are "free machining" varieties that have added sulfur, which improves their machinability. The added sulfur, however, leads to pitting corrosion, hinders weldability, and also hinders corrosion resistance due to the formation of manganese sulfide inclusions in the metal.

1.2.4.1 *Austenitic "300 Series"*

The austenitic family of stainless steels is nonmagnetic, and while they may be work hardened, they are not hardenable through heat treating. In their soft state, they are highly ductile and have elongation yields of 60 percent. This enables austenitic stainless steels to be deep drawn into kitchen sinks and laundry tubs. Due to their high levels of chromium and nickel, they are the most corrosion resistant types of stainless steels and are often used in the chemical industry for piping and storage vats. An austenitic grade with an "L" suffix denotes carbon content held to less than 0.03% and is used where the steel is to be welded.

1.2.4.2 Martensitic

Martensitic stainless steels are hardenable through heat treating, which, combined with their corrosion resistance, make them ideal materials for cutlery, scissors, and surgical instruments, where a keen, clean edge is needed. These steels may have over 100 points (1 percent) of carbon, which allows their deep hardening qualities. These hard, corrosion-resistant steels are also valuable for wear plates, ball bearings, strainers, springs, and other components used in corrosive environments. In these cases, martensitic alloys combine the best of both worlds from carbon and stainless steels. Many of the series 400 stainless steels are martensitic.

1.2.4.3 Ferritic

Ferritic stainless steels are magnetic but not heat treatable. They have been produced to combat stress corrosion cracking (SCC), which can plague other grades of stainless steel in certain atmospheric and mechanical situations. With regard to straight corrosion resistance, ferritic steels fall in between austenitic and martensitic varieties. Their resistance to SCC allows them to be used in gasoline and diesel engine exhaust systems, where the use of other stainless steels would fail. Other popular applications include furnace and heater parts and heat exchangers. Not related to its resistance to SCC, ferritic stainless steels are also used for architectural members and trim items; this is due to their lower cost as compared to other stainless steels.

1.2.4.4 Type 304 (UNS S30400), Type 304L (UNS S30403), Type 304H (UNS S30409), and Type 303 Stainless Steels (UNS S30300)

The austenitic grade type 304 is the most common stainless steel used, due to its all-around properties. It is also known as "18/8," designating 18 percent chromium and 8 percent nickel. Like other austenitic grades, it is nonmagnetic when annealed but can develop weak attraction when sufficiently cold worked. It is not hardenable through heat treating. Type 304 is easily deep drawn into products like pot stocks, sinks, and coffee urns. Its high chromium content gives it excellent corrosion resistance where it is used in food and beverage processing equipment, marine hardware, and chemical vats. The chromium also allows lustrous finishes on flatware and general kitchen utensils.

Type 304 is not easily machined, but with the addition of sulfur, a free machining version (Type 303) is produced, enabling it to be satisfactorily milled and lathe turned. Type 303 has slightly inferior corrosion resistance, however, as compared to type 304.

Type 304 stainless has a typical carbon content of 0.08 percent, which can lead to difficulties in welding. A low carbon version (type 304L) has been produced that remedies this, as it holds its carbon content to a maximum of 0.03 percent. A high carbon version (type 304H) has been produced that has improved high-temperature properties over type 304.

Table 1.19 Properties of Stainless Steel (Type 304)[*]

	Metric	English
Density	8.00 g/cm^3	0.289 lb/in^3
Tensile Strength, Ultimate	505 MPa	73,200 psi
Tensile Strength, Yield	215 MPa	31,200 psi
Elongation at Break	70.0%	70.0%
Hardness Rockwell B	70	70

Chemical Composition

	Metric	English
Carbon, C	≤0.080%	≤0.080%
Chromium, Cr	18.0–20.0%	18.0–20.0%
Iron, Fe	66.345–74.0%	66.345–74.0%
Manganese, Mn	≤2.0%	≤2.0%
Nickel, Ni	8.0–10.5%	8.0–10.5%
Phosphorous, P	≤0.045%	≤0.045%
Silicon, Si	≤1.0%	≤1.0%
Sulfur, S	≤0.030%	≤0.030%

[*]Selected data from the MatWeb search database reprinted with permission from MatWeb, www.matweb.com.

1.2.4.5 Type 316 (UNS S31600), Type 316L (UNS S31603), and Type 316H Stainless Steels (UNS S31609)

Type 316 stainless steel differs chemically from type 304, with the addition of molybdenum (2.0 to 3.0 percent) and a higher nickel content. These differences, when compared to type 304, give type 316 better corrosion resistance in marine and chlorine environments, also resulting in a stronger stainless steel. The many uses of this metal include chemical storage vats, marine hardware, and structural items as well as food preparation equipment. Type 316 steels are also biocompatible, and supplement the vast inventory of implantable materials. As with type 304, type 316 has a low carbon version (type 316L) that is easier to weld. Type 316 is also available in a high carbon variety, type 316H, which has better high-temperature corrosion resistance properties.

Table 1.20   Properties of Stainless Steel (Type 316)[*]

	Metric	English
Density	8.00 g/cm^3	0.289 lb/in^3
Tensile Strength, Ultimate	580 MPa	84100 psi
Tensile Strength, Yield	290 MPa	42,100 psi
Elongation at Break	50.0%	50.0%
Hardness Rockwell B	79	79

Chemical Composition

Carbon, C	≤0.080%	≤0.080%
Chromium, Cr	16–18%	16–18%
Iron, Fe	61.8–72.0%	61.8–72.0%
Manganese, Mn	≤2.00%	≤2.00%
Molybdenum, Mo	2.00–3.0%	2.00–3.0%
Nickel, Ni	10.0–14.0%	10.0–14.0%
Phosphorous, P	≤0.0450%	≤0.0450%
Silicon, Si	≤1.00%	≤1.00%
Sulfur, S	≤0.0300%	≤0.0300%

[*]Selected data from the MatWeb search database reprinted with permission from MatWeb, www.matweb.com.

1.2.4.6 Type 409 Stainless Steel (UNS S40900)

Type 409 is generally the least expensive stainless steel and was specifically designed for automotive exhaust systems. This alloy falls into the ferritic family of stainless steels and, like others in its family, it is ferromagnetic, high in chromium, and not heat treatable. Type 409, unlike other ferritic stainless steels, has titanium added that helps stabilize the metal when it is welded, keeping its hardness uniform. The titanium, however, tends to streak the metal to the point of visibility, making type 409 unsuitable where appearance is vital. For its intended use, however, it excels. In the corrosive combination of heat and gasses generated in an automotive exhaust system, it outlasts carbon steel by a large factor and, like the low carbon steel that it replaced, it is easily formed and welded. Other uses of type 409 include heat exchanger tubing, culvert pipes, and automotive fuel filters. This metal is not generally milled or lathe turned, and its machinability is rated at 50 percent of AISI 1212.

Table 1.21   Properties of Stainless Steel (Type 409)[*]

	Metric	English
Density	7.80 g/cm^3	0.282 lb/in^3
Tensile Strength, Ultimate	450 MPa	65,300 psi
Tensile Strength, Yield	240 MPa	34,800 psi
Elongation at Break	25.0%	25.0%
Hardness Rockwell B	75	75

Chemical Composition

	Metric	English
Carbon, C	≤0.080%	≤0.080%

Table 1.21   Properties of Stainless Steel (Type 409)[*] *(cont.)*

	Metric	English
Chromium, Cr	11.13%	11.13%
Iron, Fe	86.0%	86.0%
Manganese, Mn	≤1.0%	≤1.0%
Phosphorous, P	≤0.045%	≤0.045%
Silicon, Si	≤1.0%	≤1.0%
Sulfur, S	≤0.045%	≤0.045%
Titanium, Ti	0.75%	0.75%

[*]Selected data from the MatWeb search database reprinted with permission from MatWeb, www.matweb.com.

1.2.4.7 *Type 430 (UNS S43000) and Type 430F (UNS S43020)*

Type 430 stainless is a ferritic, high chromium, low carbon steel. It is easily cold worked and deep drawn, and does not tend to work harden as fast as other stainless steels do. This quality allows more operations between annealings, if they are required at all. Type 430 is a relatively low-cost stainless and is used in many domestic applications including appliance trim, sinks, range hoods, and flatware. It has a very pleasing tone and takes a very good finish and polish. As with all stainless steels, the higher the finish, the more corrosion resistance it will have, as a result of the lack of crevices in which corrosive materials can hide. Type 430 is difficult to machine but is available as a sulfur added alloy (type 430F) that permits easier milling and lathe turning. Type 430F is available in solid rod form and is used to make pump components, valves, and other items needing limited corrosion protection.

Table 1.22   Properties of Stainless Steel (Type 430)[*]

	Metric	English
Density	7.80 g/cm^3	0.282 lb/in^3
Tensile Strength, Ultimate	538 MPa	78,000 psi
Tensile Strength, Yield	441 MPa	64,000 psi
Elongation at Break	32.0%	32.0%
Chemical Composition		
Carbon, C	≤0.12%	≤0.12%
Chromium, Cr	11.0%	11.0%
Iron, Fe	87.0%	87.0%
Manganese, Mn	≤1.0%	≤1.0%
Phosphorous, P	≤0.040%	≤0.040%
Silicon, Si	≤1.0%	≤1.0%
Sulfur, S	≤0.030%	≤0.030%

[*]Selected data from the MatWeb search database reprinted with permission from MatWeb, www.matweb.com.

1.2.4.8 *Type 440C Stainless Steels (UNS S44004)*

Type 440C stainless steel is a martensitic variety noted for its wear resistance and high strength. It is a high carbon, high chromium, nickel free stainless tool steel capable of intense hardening, up to RC 60, and moderate corrosion resistance in mildly active environments. Type 440C, due to its wear resistance properties, is used in cutlery ranging from hair scissors to hunting and kitchen knives. It can hold a keen edge and can take a mirror polish—two qualities needed for fine surgical instruments. Type 440C is also used in roller and ball bearings, raceways, plain bearings, and slides. Its mechanical properties are comparable to a general-purpose tool steel, such as AISI 01, but with the benefit of corrosion resistance. Type 440C can be successfully welded but, as with other high carbon steels, preheating and annealing must be accomplished beforehand and afterwards, respectively.

Its machinability is average for a tool steel (40 percent of AISI 1212), but of course, machining must be performed when the steel is in its annealed state. For easier machining, type 440F has been formulated with added sulfur that increases machinability up to 50 percent of AISI 1212.

In the 440 family, there are also the "A" and "B" designations. Type 440A has a lower carbon content than 440B, which itself is lower in carbon, and hence hardenability, than 440C. Neither 440A nor 440B steels are very popular, but with their lower carbon content they would be slightly more ductile, in sacrifice for their slight lack of hardness and strength when compared to the "C" steel.

Table 1.23 Properties of Stainless Steel (Type 440C)[*]

	Metric	English
Density	7.80 g/cm^3	0.282 lb/in^3
Tensile Strength, Ultimate	760–1970 MPa	11,0000–286,000 psi
Tensile Strength, Yield	450–1900 MPa	65,300–276,000 psi

Table 1.23 Properties of Stainless Steel (Type 440C)[*] *(cont.)*

	Metric	English
Elongation at Break	2.00–14.0%	2.00–14.0%
Hardness Rockwell B	97.0	97.0

Chemical Composition

	Metric	English
Carbon, C	1.10%	1.10%
Chromium, Cr	17.0%	17.0%
Iron, Fe	79.15%	79.15%
Mn	1.00%	1.00%
Molybdenum, Mo	0.750%	0.750%
Silicon, Si	1.00%	1.00%

[*]Selected data from the MatWeb search database reprinted with permission from MatWeb, www.matweb.com.

2
Nonferrous Alloys

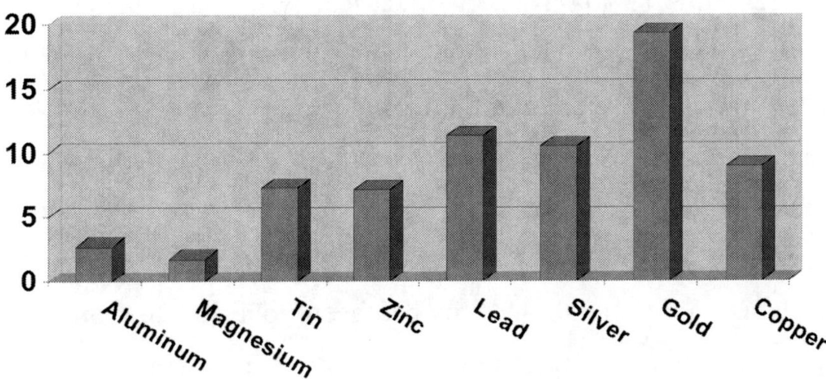

FIGURE 2.1 Densities, in g/cm^3, of various metals.

2.1 COPPER AND ITS ALLOYS

Copper (atomic number 29, symbol Cu) is one of our most important metallic elements, with a domestic U.S. consumption of over 2.1 million metric tons in 2007, according to the United States Geological Survey. It is the mother of hundreds of alloys, including the brasses and bronzes. Copper has excellent thermal and electrical conductivity and is second only to silver in these regards. It is a soft metal and both extremely ductile and malleable. It has an orange-peach color when pure and unoxidized but, when left to weather, it takes on a green patina coating that protects the unexposed copper underneath. The Statue of Liberty is a prime example of this, as her exterior is made of pure copper sheeting, which has remained exposed in the salt air of New York Harbor for over 120 years.

With its excellent ductility and electrical conductivity, pure copper finds its main use as electrical wire. Its high thermal conductivity allows it to quickly dissipate heat when used in heat exchangers including automotive engine radiators. Its corrosion resistance properties make it useful as potable water pipes, valves, and fittings. It is easily soldered and brazed, and it can be welded with a gas shielded arc. Copper is used worldwide as a coin metal, and this has been done for thousands of years. It can be work formed by all common methods, including shearing, punching, spinning, and drawing, but its properties make it difficult to machine.

As copper is cold worked, its tensile and yield strength is increased considerably—so much, in fact, that a fully work hardened copper can have twice its original tensile strength, and its yield can be increased more than fourfold. This hardened metal can be returned to its original softness by annealing, where controlled heating and cooling will recrystallize and de-stress the metal at the atomic level. In instances where work hardened copper needs to remain hard at higher temperatures, anneal-resistant coppers have been formulated.

2.1.1 C10100 Oxygen-Free Electronic Copper

Alloy 101 is designated as 99.99 percent pure—the highest purity of any commercially available copper, containing a maximum oxygen content of five parts per million. As this alloy has the best electrical and thermal conductivity of any copper alloy, it shows that when additional elements are added to copper, its conductivity suffers. Its low oxygen content protects against hydrogen embrittlement, where the metal looses its tensile ductility. Alloy 101 copper is easily hot worked (extruding and forging) and cold worked (deep drawing and bending).

It is used for electrical buss bars, conductors, and contacts where the utmost in conductivity and reliability are required. Like all of the straight coppers, it is difficult to machine and has a machinability rating of 20 percent as compared to alloy 360 leaded brass.

Table 2.1   Properties of Oxygen-Free Electronic Copper

	Metric	English
Density	8.89–8.94 g/cm^3	0.321–0.323 lb/in^3
Melting Point	1083°C	1981°F
Yield Strength	69.0–365 MPa	10,000–52,900 psi
Machinability	20%	20%

Chemical Composition		
Copper, Cu	≥99.99%	≥99.99%

2.1.2 Copper Alloy C11000 Electrolytic Touch Pitch Copper

Alloy C11000 is the most widely used member of the straight coppers. Having a purity level of 99.90 percent, it is suitable for all but the most exacting applications. One drawback to this alloy is its small (0.04 percent) amount of oxygen present. Even with this tiny percentage, hydrogen embrittlement can occur when the alloy is heated in a reducing atmosphere, such as during gas welding. In hydrogen embrittlement, hydrogen ions couple with the oxygen in the copper, forming water vapor under pressure. This vapor leaves pits and pores in the copper, effectively turning it brittle. Although some types of welding can be accomplished with this metal, oxygen-free copper alloys are available that reduce the risk of embrittlement to zero.

Table 2.2 PROPERTIES OF COPPER ALLOY C11000

	Metric	English
Density	8.89 g/cm^3	0.321 lb/in^3
Melting Point	1065–1083°C	1949–1981°F
Thermal Conductivity	388 W/m·K	2690 Btu-in/hr·ft^2·°F
Yield Strength	76.0 MPa	11000 psi
Machinability	20%	20%
Chemical Composition		
Copper, Cu	99.9%	99.9%
Oxygen, O	0.040%	0.040%

2.1.3 C12200 Phosphorus-Deoxidized Copper

The deoxidation process used in this alloy allows it to be successfully welded and brazed without the risk of embrittlement, at a lower cost than alloy C10100. The main use of this alloy is for copper tubing and pipe, but it is also available as plate and sheeting. Alloy C12200 has better ductility than alloy C11000, allowing fewer stages in progressive deep drawing operations.

Table 2.3   Properties of C12200 Copper[*]

	Metric	English
Density	8.94 g/cm^3	0.323 lb/in^3
Yield Strength	69.0–345 MPa	10,000–50,000 psi
Machinability	20%	20%
Chemical Composition		
Copper, Cu	99.9%	99.9%
Phosphorous, P	0.020%	0.020%

[*]Selected data reprinted with permission from the Copper Development Association (CDA), www.copper.org/resources/properties/.

2.1.4 C14500 Tellurium Copper

Adding a small amount of tellurium to pure copper does not effect its electrical conductivity to a great extent, but it enhances its machinability from a rating of 20 to 85 percent of that of alloy C36000 brass. Nearly pure copper that machines like free cutting brass is useful for lathe turned hardware, including threaded studs and bolts, that must carry current without appreciable loss. The tellurium also has a deoxidizing effect, preventing embrittlement when this alloy is heated in a reducing atmosphere.

Table 2.4   Properties of C14500 Tellurium Copper[*]

	Metric	English
Density	8.94 g/cm^3	0.323 lb/in^3
Melting Point	1080°C	1976°F
Electrical Conductivity	0.544 MS/cm	93% IACS
Thermal Conductivity	355 W/m·K	2460 Btu-in/hr·ft^2·°F
Yield Strength	10,000 psi	69.0 MPa
Machinability	85%	85%

Chemical Composition

	Metric	English
Copper, Cu	99.5%	99.5%
Phosphorous, P	0.0040–0.012%	0.0040–0.012%
Tellurium, Te	0.40–0.60%	0.40–0.60%

[*]Selected data reprinted with permission from the Copper Development Association (CDA), www.copper.org/resources/properties/.

2.1.5 C15000 Zirconium Copper

Zirconium copper offers useful properties not found together in other coppers, such as strength and electrical conductivity. While conductive leaf springs and contacts made of Beryllium copper may only conduct electricity at 48% IACS, the same leaf can be made with alloy 150 zirconium that possesses 93% IACS conductivity. Unlike pure copper, zirconium copper has an elevated softening temperature, allowing it to be used in conditions of higher heat while maintaining its hardness.

Table 2.5 Properties of C15000 Zirconium Copper[*]

	Metric	English
Density	8.89 g/cm^3	0.321 lb/in^3
Melting Point	980–1080°C	1800–1980°F
Thermal Conductivity	367 W/m·K	2550 Btu-in/hr·ft^2·°F
Hardness, Rockwell B	72	72
Tensile Strength, Ultimate	430 MPa	62,400 psi
Tensile Strength, Yield	385 MPa	55,800 psi
Elongation at Break	8.00%	8.00% in 50 mm
Modulus of Elasticity	129 GPa	18,700 ksi
Machinability	20%	20%

Chemical Composition

	Metric	English
Copper, Cu	99.85%	99.85%
Zirconium, Zr	0.13–0.20%	0.13–0.20%

[*]Selected data reprinted with permission from the Copper Development Association (CDA), www.copper.org/resources/properties/.

2.1.6 Beryllium Coppers

Beryllium coppers possess the highest strength of any copper alloys, approaching that of mild steel. However, being nonferrous, they are nonmagnetic, non-sparking, and corrosion resistant. Beryllium coppers can be precipitously hardened as well as cold work hardened. The dust and fumes of beryllium copper can cause serious pulmonary illness when inhaled. Safety concerns (including using filtering devices and air quality measurement) should be kept in mind with certain operations, including abrasive cutting, grinding, and welding. General handling, most machining operations, and cold working, however, do not present a hazard. Two families of beryllium copper exist: high beryllium and low beryllium.

High beryllium coppers (alloys C17000, C17200, and C17300) are the hardest varieties, with 1.7 to 2.0 percent beryllium added. Typical uses include non-sparking tools for use in explosive atmospheres, coil and leaf springs, and other components that rely on their high hardness. Alloy C17300 goes one step further by including lead, which gives it a free machining quality. Beryllium copper also possesses peculiar tonal qualities that make it useful in tambourines and other instruments. Some golfers also make use of this alloy in their putters.

Low beryllium alloys (C17500 and C17510) have better electrical conductivity than the high alloys, due to their lower amounts of beryllium added. This small amount of 0.2 to 0.7 percent is still sufficient, however, to add great strength to the otherwise soft copper. Small amounts of cobalt and nickel are also added to these alloys to improve their hardness and strength. Uses include electrical contacts and other conductive components.

Table 2.6   Properties of C17200 Beryllium Copper[*]

	Metric	English
Density	8.25 g/cm^3	0.298 lb/in^3
Melting Point	865–980°C	1590–1800°F
Thermal Conductivity	105–130 W/m·K	729–902 Btu-in/hr·ft^2·°F
Machinability	20%	20%

Table 2.6   Properties of C17200 Beryllium Copper[*] *(cont.)*

	Metric	English
Chemical Composition		
Copper	98.1%	98.1%
Beryllium	1.9%	1.9%

[*]Selected data reprinted with permission from the Copper Development Association (CDA), www.copper.org/resources/properties/.

Table 2.7   C17300 Leaded Beryllium Copper[*]

	Metric	English
Density	8.25 g/cm^3 @ 20°C	0.298 lb/in^3 @ 68°F
Melting Point	982°C	1800°F
Electrical Conductivity	129 MS/cm @ 20°C	22% IACS @ 68°F
Thermal Conductivity	107.3 W/m·K @ 20°C	62 Btu·ft/(hr·ft^2·°F) @ 68°F
Hardness (round section, TD04 Temper)	Rockwell B 95	Rockwell B 95
Yield Strength	724 MPa	105 ksi
Machinability	50%	50%
Chemical Composition		
Copper, Cu	97.7%	97.7%
Beryllium, Be.	1.9%	1.9%
Lead, Pb	0.4%	0.4%

[*]Selected data reprinted with permission from the Copper Development Association (CDA), www.copper.org/resources/properties/.

Table 2.8   Properties of C17500 Beryllium Copper[*]

	Metric	English
Density	8.61 g/cm^3 @ 20°C	0.311 lb/in^3 @ 68°F
Melting Point	1068°C	1955°F
Electrical Conductivity	0.263 MS/cm @ 20°C	45%IACS @ 68°F
Thermal Conductivity	207.7 W/m·K @ 20°C	120 Btu·ft/(hr·ft^2·°F) @ 68°F
Hardness (round section, TD04 temper)	Rockwell B 68	Rockwell B 68
Yield Strength	496 MPa	72 ksi
Machinability	40%	40%

Chemical Composition

	Metric	English
Copper, Cu	96.9%	96.9%
Beryllium, Be	0.55%	0.55%
Cobalt, Co	2.6%	2.6%

[*]Selected data reprinted with permission from the Copper Development Association (CDA), www.copper.org/resources/properties/.

Table 2.9   Properties of C17510 Beryllium Copper[*]

	Metric	English
Density	8.77 g/cm^3 @ 20°C	317 lb/in^3 @ 68°F
Melting Point	1068°C	1955°F
Electrical Conductivity	0.281 MS/cm @ 20°C	48% IACS @ 68°F
Thermal Conductivity	207.7 W/m·K @ 20°C	120 Btu·ft/ (hr·ft 2·°F) @ 68°F

Table 2.9 Properties of C17510 Beryllium Copper * *(cont.)*

	Metric	English
Hardness (round section, TD04 temper)	Rockwell B 68	Rockwell B 68
Yield Strength	496 MPa	72 ksi
Machinability	40%	40%

Chemical Composition		
Copper, Cu	97.8%	97.8%
Beryllium	0.4%	0.4%
Nickel	1.8%	1.8%

*Selected data reprinted with permission from the Copper Development Association (CDA), www.copper.org/resources/properties/.

2.1.7 The Brasses and Bronzes

In general, brass is an alloy made of copper and zinc, with copper being the primary metal. Bronze, on the other hand, is copper alloyed with tin. These names, however, are at times applied to alloys according to their color more than their elemental makeup, and there are actual bronzes commonly known as brasses and vice versa. Some alloys also blur the line by having both tin and zinc included in their makeup. Several alloys also include trace elements in varying proportions to secure specific properties. Adding 2 or 3 percent of lead by weight to a standard brass makes the best machining alloy of all metals (Fig. 2.2). A small amount of phosphor added to a straight bronze dramatically increases its hardness, strength, and resistance to wear. The single greatest property of the brasses and

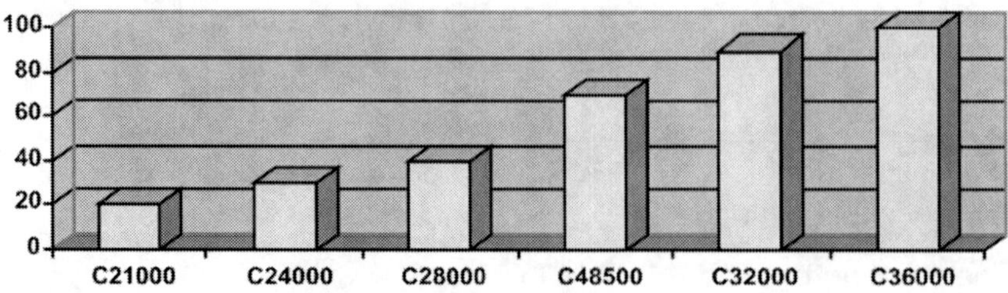

FIGURE 2.2 Relative machinability of brass alloys (%).

bronzes, however, is their corrosion resistance. There are alloys that can withstand salt or chemical atmospheres that would reduce steel and iron to rubble, and look good doing so. Brass has a very appealing golden yellow color, and both it and the deeper-shaded bronze can be polished to a bright mirror finish. They are both used decoratively for trim, architectural members, and exposed hardware items.

2.1.7.1 C21000 Gilding Metal

This copper alloy is produced with a 95/5 makeup with zinc as its secondary element. It offers good strength and excellent corrosion resistance, and it is easily deep drawn. As with most straight copper zinc brasses, it is very difficult to machine but can be cold worked with excellent results. It is easily soldered and brazed. One final property is its resistance to stress cracking corrosion, where normally soft and ductile metals suffer cracking over time when subject to a corrosive atmosphere. This problem increases proportionately as the zinc content in brass increases.

One of its primary uses is as a base metal for gold plating on metals and medallions, hence its name "gilding metal." This alloy is also used extensively in the ordnance field for bullet jackets, fuse caps, and primers.

Table 2.10   Properties of C21000 Gilding Metal[*]

	Metric	English
Density	8.86 g/cm^3 @ 20°C	0.320 lb/in^3 @ 68°F
Melting Point	1066°C	1950°F
Electrical Conductivity	0.328 MS/cm @ 20°C	56% IACS @ 68°F
Thermal Conductivity	233.6 W/m·K @ 20°C	135 Btu·ft/ (hr·ft^2·°F) @ 68°F
Hardness (flat section, H02 temper)	Rockwell B 52	Rockwell B 52
Yield Strength	276 MPa	40 ksi
Machinability	20%	20%

[*]Selected data reprinted with permission from the Copper Development Association (CDA), www.copper.org/resources/properties/.

2.1.7.2 C22000 Commercial Bronze

Although technically a 90/10 brass, it is commonly referred to as a bronze due to its color. As such, alloy C22000 is often used for decorative items including exposed rivets and bolts on architectural and structural members, as well as ornamental trim and costume jewelry. But besides good looks, alloy C22000 exhibits moderate strength, good corrosion resistance, and formability.

Table 2.11   Properties of C22000 Bronze[*]

	Metric	English
Density	8.8 g/cm^3 @ 20°C	318 lb/in^3 @ 68°F
Melting Point	1043°C	1910°F

Table 2.11   Properties of C22000 Bronze[*] *(cont.)*

	Metric	English
Electrical Conductivity	0.257 MS/cm @ 20°C	44%IACS @ 68°F
Thermal Conductivity	188.7 W/m·K @ 20°C	109 Btu·ft/(hr·ft^2·°F) @ 68°F
Hardness (flat section, H02 temper)	Rockwell B 58	Rockwell B 58
Yield Strength	310 MPa	45 ksi
Machinability	20%	20%

Chemical Composition

	Metric	English
Copper, Cu	90%	90%
Zinc, Zn	10%	10%

[*]Selected data reprinted with permission from the Copper Development Association (CDA), www.copper.org/resources/properties/.

2.1.7.3 C24000 Low Brass

This 80/20 brass has very good strength and corrosion resistance; together with its excellent cold and hot working capabilities, it makes a very common material for hardware, pump components, and rotor bars in electrical motors. Due to its very pleasing yellow color, it is also used in ornamental objects, jewelry items, and musical instruments.

Table 2.12   Properties of C2400 Low Brass[*]

	Metric	English
Density	8.66 g/cm^3 @ 20°C	313 lb/in^3 @ 68°F
Melting Point	999°C	1830°F

Table 2.12   Properties of C2400 Low Brass [*] **(cont.)**

	Metric	English
Electrical Conductivity	0.187 MS/cm @ 20 C	32%IACS @ 68°F
Thermal Conductivity	140.2 W/m·K @ 20 C	81 Btu·ft/ (hr·ft^2·°F) @ 68°F
Hardness (flat section, H02 temper)	Rockwell B 70	Rockwell B 70
Yield Strength	345 MPa	50 ksi
Machinability	30%	30%

Chemical Composition		
Copper, Cu	80%	80%
Zinc, Zn	20%	20%

[*]Selected data reprinted with permission from the Copper Development Association (CDA), www.copper.org/resources/properties/.

2.1.7.4 C26000 Cartridge Brass

Consisting of a simple alloy of 70 percent copper and 30 percent zinc, cartridge brass has the highest ductility of any brass alloy, making it perfect for not only forming the ammunition cartridge, but also for absorbing the internal explosion of the cartridge without cracking or becoming brittle. Besides its ductility, cartridge brass has good hardness and strength, surpassing many other brasses in both categories. Alloy 260 is, however, subject to stress crack corrosion, and ammunition formed from it must not be stored in ammonia or acidic atmospheres.

Alloy 260 brass is available in sheets, tubes, and solid sections. It solders and brazes very well but is difficult to machine as compared to a leaded brass. It can be cold worked very easily, making it a favorite for deep drawing and spinning.

Table 2.13   Properties of C26000 Cartridge Brass[*]

	Metric	English
Density	8.53 g/cm^3 @ 20°C	0.308 lb/in^3 @ 68°F
Melting Point	954°C	1750°F
Electrical Conductivity	0.164 MS/cm @ 20°C	28% IACS @ 68°F
Thermal Conductivity	121.2 W/m·K @ 20°C	70 Btu·ft/(hr·ft^2·°F) @ 68°F
Hardness (flat section, H04 temper)	Rockwell B 82	Rockwell B 82
Yield Strength	434 MPa	63 ski
Machinability	30%	30%

Chemical Composition (Nominal)

Copper, Cu	70%	70%
Zinc, Zn	30%	30%

[*]Selected data reprinted with permission from the Copper Development Association (CDA), www.copper.org/resources/properties/.

2.1.7.5 *C28000 Muntz Metal*

Alloy C28000, or Muntz metal, was named after its developer. Originally used as anti-fouling panels on ships, this alloy is still in demand today for architectural and structural panels due to its combination of strength, appealing color, and malleability. Other uses include decorative hardware, marine fasteners, and hot forgings.

Table 2.14 Properties of C28000 Muntz Metal[*]

	Metric	English
Density	8.44 g/cm^3 @ 20°C	0.303 lb/in^3 @ 68°F
Melting Point	899°C	1650°F
Electrical Conductivity	0.164 MS/cm @ 20°C	28% IACS @ 68°F
Thermal Conductivity	123 W/m·K @ 20°C	71 Btu·ft/(hr·ft^2·°F) @ 68°F
Hardness (flat section, H02 temper)	Rockwell B 75	Rockwell B 75
Yield Strength	438 MPa	50 ksi
Machinability	40%	40%
Chemical Composition		
Copper, Cu	60%	60%
Zinc, Zn	40%	40%

[*]Selected data reprinted with permission from the Copper Development Association (CDA), www.copper.org/resources/properties/.

2.1.7.6 C32000 Leaded Red Brass

The addition of lead to this 85/13 brass dramatically increases machinability up to 90 percent of C36000 free cutting brass, while maintaining a much higher degree of strength. It is easily soldered and brazed, but the lead content makes it difficult to weld. Available in round and rectangular bar stock, alloy C32000 brass is mainly used for machined hardware and components that need its high strength and corrosion resistance.

Table 2.15   Properties of C32000 Leaded Brass[*]

	Metric	English
Density	8.75 g/cm^3 @ 20°C	0.316 lb/in^3 @ 68°F
Melting Point	1024°C	1875°F
Electrical Conductivity	0.211 MS/cm @ 20°C	36%IACS @ 68°F
Thermal Conductivity	155.8 W/m·K @ 20°C	90 Btu·ft/(hr·ft^2·°F) @ 68°F
Hardness (round section, H02 temper)	Rockwell B 75	Rockwell B 75
Yield Strength	414 MPa	60 ksi
Machinability	90%	90%

Chemical Composition (Nominal)

	Metric	English
Copper, Cu	85%	85%
Zinc, Zn	13.2%	13.2%
Lead, Pb	1.8%	1.8%

[*]Selected data reprinted with permission from the Copper Development Association (CDA), www.copper.org/resources/properties/.

2.1.7.7 C36000 Free Cutting Brass

The benchmark of machinability, alloy 360 leaded brass (Figs. 2.3, 2.4) is by far the most widely used alloy of the brasses and bronzes for machined parts. The addition of 3.1 percent lead to a mixture of 61.5 percent copper and 35.4 percent zinc produces this alloy. It is favored for not only its superb machinability but also its corrosion resistance, allowing parts to be produced at lower costs than can be achieved with plated steel when used in corrosive atmospheres. The addition of lead is twofold; first, it

FIGURE 2.3 Free Cutting Brass (Alloy UNS C36000) Shapes.

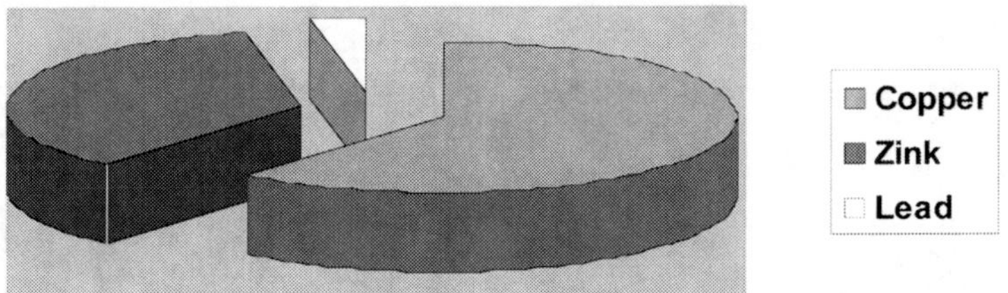

FIGURE 2.4 Elemental makeup of alloy C36000 free cutting brass.

provides lubrication between the cutting tool and the work. Second, the lead weakens the bond of the crystalline structure of the metal, allowing the crystals to sheave away from each other along their boundaries rather than being fractured as the part is machined. This allows excellent finishes, with very low tool wear, even when using high cutting speeds and feeds.

Table 2.16   Properties of C36000 Brass[*]

	Metric	English
Density	8.5 g/cm^3 @ 20°C	0.307 lb/in^3 @ 68°F
Melting Point	899°C	1650°F
Electrical Conductivity	0.152 MS/cm @ 20°C	26% IACS @ 68°F
Thermal Conductivity	116.0 W/m·K @ 20°C	67 Btu·ft/(hr·ft^2·°F) @ 68°F
Hardness (round section, H04 temper)	Rockwell B 78	Rockwell B 78

[*]Selected data reprinted with permission from the Copper Development Association (CDA), www.copper.org/resources/properties/.

2.1.7.8 C46400 Naval Brass

Known for its corrosion resistance and high strength, alloy C46400 is often used in marine environments for propeller shafting, rivets, and general hardware. Other uses include heat exchangers, boilerplates, structural members, and other applications where a salt corrosive atmosphere may be present.

Naval brass C46400 offers excellent soldering and brazing ability and can be easily hot worked, having an outstanding forgeability rating. Its machinability is poor, rated at 30 percent as compared to brass alloy C36000.

Table 2.17   Properties of C46400 Naval Brass[*]

	Metric	English
Density	8.41 g/cm^3 @ 20°C	0.304 lb/in^3 @ 68°F
Melting Point	899°C	1650°F

Table 2.17   Properties of C46400 Naval Brass* *(cont.)*

	Metric	English
Electrical Conductivity	0.152 MS/cm @ 20°C	26%IACS @ 68°F
Thermal Conductivity	116.0 W/m·K @ 20 C	67 Btu·ft/(hr·ft^2·°F) @ 68 F
Hardness (round section, H01 temper)	Rockwell B 75	Rockwell B 75
Yield Strength	276 MPa	40 ksi
Machinability	30%	30%

Chemical Composition (Nominal)

	Metric	English
Copper, Cu		
Zinc, Zn	39.3%	39.3%
Tin, Sn	0.7%	0.7%

*Selected data reprinted with permission from the Copper Development Association (CDA), www.copper.org/resources/properties/.

2.1.7.9 C48500 Naval Brass, High Leaded

Alloy C48500 is similar to its naval brass cousin, alloy C46400, in corrosion resistance and strength, but the addition to 0.7 percent lead allows easy machining for parts requiring such steps. The lead does not interfere with its hot working ability, allowing forgings of this alloy to be subsequently drilled, tapped, and so on with good results.

Table 2.18   Properties of C48500 Naval Brass*

	Metric	English
Density	8.44 g/cm^3 @ 20°C	0.305 lb/in^3 @ 68°F

Table 2.18   Properties of C48500 Naval Brass [*] **(cont.)**

	Metric	English
Melting Point	899°C	1650°F
Electrical Conductivity	0.152 MS/cm @ 20°C	26% IACS @ 68°F
Thermal Conductivity	0.152 MS/cm @ 20°C	67 Btu·ft/(hr·ft^2·°F) @ 68°F
Hardness (round section, H02 temper)	Rockwell B 82	Rockwell B 82
Yield Strength	365 MPa	53 ksi
Machinability	70%	70%

Chemical Composition (Nominal)

	Metric	English
Copper, Cu	60.5%	60.5%
Zinc, Zn	37.5%	37.5%
Lead, Pb	1.8%	1.8%
Tin, Sn	0.7%	0.7%

[*]Selected data reprinted with permission from the Copper Development Association (CDA), www.copper.org/resources/properties/.

2.1.7.10 C51000 Phosphor Bronze

Alloy C51000 is the most common alloy of phosphor bronze and has several key properties that make it useful for bearings, bushings, and clutch plates. These properties include a high degree of hardness and strength, low coefficient of friction, and a high resistance to wear. It also is easily cold worked and is readily soldered and brazed. Alloy C51000 also takes on an excellent spring temper and is used for electrical contacts and connectors, albeit it has relatively low electrical conductivity.

Table 2.19   Properties of C51000 Phosphor Bronze[*]

	Metric	English
Density	8.86 g/cm^3 @ 20°C	0.320 lb/in^3 @ 68°F
Melting Point	1049°C	920°F
Electrical Conductivity	0.088 MS/cm @ 20°C	15%IACS @ 68°F
Thermal Conductivity	69.2 W/m·K @ 20 C	40 Btu·ft/(hr·ft^2·°F) @ 68 F
Hardness (flat section, H04 temper)	Rockwell B 87	Rockwell B 87
Yield Strength	579 MPa	84 ksi
Machinability	20%	20%
Chemical Composition		
Copper, Cu	94.8%	94.8%
Tin, Sn	5.0%	5.0%
Phosphorus, P	0.2%	0.2%

[*]Selected data reprinted with permission from the Copper Development Association (CDA), www.copper.org/resources/properties/.

2.1.7.11 *C54400 High Leaded Phosphor Bronze*

The properties of alloy C54400 are similar to C51000 in its superb bearing qualities, but the addition of a small amount of lead increases its machinability dramatically for lathe-turned bushings, bearings, and shafts.

Table 2.20   Properties of C54400 Phosphor Bronze[*]

	Metric	English
Density	8.89 g/cm^3 @ 20°C	0.320 lb/in^3 @ 68°F

Table 2.20   Properties of C54400 Phosphor Bronze[*] *(cont.)*

	Metric	English
Melting Point	999°C	1830°F
Electrical Conductivity	0.111 MS/cm @ 20°C	19%IACS @ 68°F
Thermal Conductivity	86.5 W/m·K @ 20°C	50 Btu·ft/(hr·ft^2·°F) @ 68°F
Hardness (flat section, H02 temper)	Rockwell B 68	Rockwell B 68
Yield Strength	276 MPa	40 ksi
Machinability	80%	80%

Chemical Composition (Nominal)

	Metric	English
Copper, Cu	88%	88%
Tin, Sn	4.0%	4.0%
Zinc, Zn	3.0%	3.0%
Lead, Pb	3.8%	3.8%
Phosphorus, P	0.25%	0.25%

[*]Selected data reprinted with permission from the Copper Development Association (CDA), www.copper.org/resources/properties/.

2.1.7.12 C65500 High Silicon Bronze (Herculoy)

Silicon bronzes are some of the most corrosion resistant copper alloys available, and they also offer good strength and hardness. They are both hot and cold worked with good results but will work harden very rapidly, requiring frequent annealings. Typical uses include marine hardware, pump and motor shafting, and valve components. Alloy C65500 can be welded, brazed, and soldered with good results, and as such are fabricated as chemical containment vessels and piping.

Table 2.21 Properties of C65500 High Silicon Bronze[*]

	Metric	English
Density	8.53 g/cm^3 @ 20°C	0.308 lb/in^3 @ 68
Melting Point	1027°C	1880°F
Electrical Conductivity	36.3 W/m·K @ 20°C	21 Btu·ft/(hr·ft^2·°F) @ 68°F
Thermal Conductivity	36.3 W/m·K @ 20°C	21 Btu·ft/(hr·ft^2·°F) @ 68°F
Hardness (flat section, H06 temper)	Rockwell B 95	Rockwell B 95
Yield Strength	414 MPa	60 ksi
Machinability	30%	30%
Chemical Composition		
Copper, Cu	97%	97%
Silicon, Si	3.0%	3.0%
Molybdenum, Mo	1%	1%

[*]Selected data reprinted with permission from the Copper Development Association (CDA), www.copper.org/resources/properties/.

2.1.7.13 C69300 ECO-Brass

Alloy C69300 ECO-Brass is a non-leaded replacement for alloy C36000 free cutting brass. Useful in potable water system valves, fittings and faucets, it is slowly replacing leaded brasses especially in areas where strict laws prohibit their use in residential and commercial fixtures. Other uses include screw machine products and as a general material for the hobby machinist. Alloy C69300 is weldable, can be soldered and brazed, and has a high degree of strength.

Table 2.22   Properties of *C69300 ECO-Brass*[*]

	Metric	English
Density	8.3 g/cm^3 @ 20°C	0.300 lb/in^3 @ 68°F
Melting Point	880°C	1616°F
Electrical Conductivity	0.046 MS/cm @ 20°C	8%IACS @ 68°F
Thermal Conductivity	68F37.7 W/m·K @ 20°C	21.80 Btu·ft/(hr·ft^2·°F) @ 68°F
Hardness (round section, H02 temper)	Rockwell B 80	Rockwell B 80
Yield Strength	338 MPa	49 ksi
Machinability	70%	70%

Chemical Composition (Nominal)

	Metric	English
Copper, Cu	75%	75%
Zinc, Zn	21.9%	21.9%
Silicon, Si	3.0%	3.0%
Phosphorus, P	0.10%	0.10%

[*]Selected data reprinted with permission from the Copper Development Association (CDA), www.copper.org/resources/properties/.

2.1.7.14 C71500 Cupronickel

Cupronickel alloy C71500 is used extensively for boiler and saltwater tubes. This alloy can safely withstand the combination of high pressure, temperature, corrosive liquids, and turbulence. It is readily welded, soldered, and brazed and can be both hot and cold worked with good results. Other uses for alloy C71500 include general hardware components that are subject to extreme conditions.

The US nickel coins are made from cupronickel alloy C71300, consisting of a 75/25 copper nickel blend. Although containing only 25 percent nickel, it completely overtakes the orange/peach color of the predominant copper.

Table 2.23   Properties of C71500 Cupronickel[*]

	Metric	English
Density	8.94 g/cm^3 @ 20°C	0.323 lb/in^3 @ 68°F
Melting Point	1238°C	2260°F
Electrical Conductivity	0.027 MS/cm @ 20°C	4.60%IACS @ 68°F
Thermal Conductivity	29.4 W/m·K @ 20°C	17 Btu·ft/(hr·ft^2·°F) @ 68°F
Hardness (round section, H02 temper)	Rockwell B 80	Rockwell B 80
Yield Strength	517 MPa	75 ksi
Machinability	20%	20%

Chemical Composition (Nominal)

	Metric	English
Copper, Cu	69.5%	69.5%
Nickel, Ni	30%	30%
Iron, Fe	0.5%	0.5%

[*]Selected data reprinted with permission from the Copper Development Association (CDA), www.copper.org/resources/properties/.

2.1.7.15 C73500 Nickel Silver, German Silver

Nickel silver contains no silver (Fig. 2.5), but its appearance resembles the precious metal. Alloy C73500 is used mainly ornamentally for jewelry, hair clips, cigarette cases, and similar items.

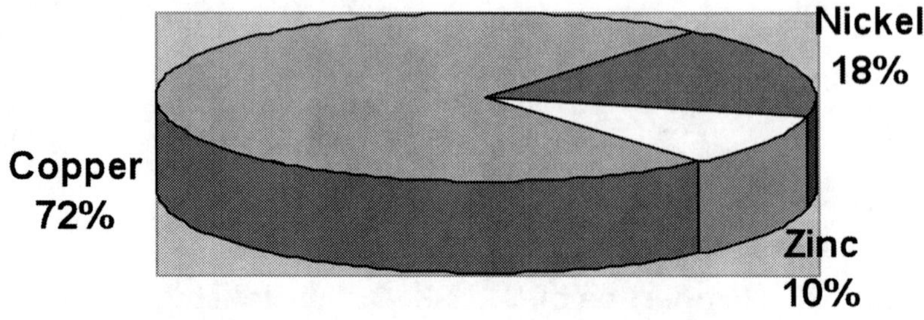

FIGURE 2.5 Elemental makeup of German silver.

It can be readily welded, brazed, and soldered, and it can take on a high polish that doesn't readily oxidize. It is very malleable and can be worked into thin sheets and ribbons to be punched to shape.

Alloy C73500 also has industrial uses as marine hardware, taking advantage of not only its pleasing appearance but also its corrosion resistance properties and high strength.

Table 2.24   Properties of C73500 Nickel Silver[*]

	Metric	English
Density	8.8 g/cm^3 @ 20°C	0.318 lb/in^3 @ 68°F
Melting Point	1135°C	2075°F
Electrical Conductivity	0.038 MS/cm @ 20°C	6.50% IACS @ 68°F
Thermal Conductivity	36.4 W/m·K @ 20°C	21 Btu·ft/(hr·ft^2·°F) @ 68°F
Hardness (flat section, H04 temper)	Rockwell B 80	Rockwell B 80
Yield Strength	510 MPa	74 ksi
Chemical Composition (Nominal)		
Copper, Cu	72%	72%

Table 2.24   Properties of C73500 Nickel Silver* *(cont.)*

	Metric	English
Nickel, Ni	18%	18%
Zinc, Zn	10%	10%

*Selected data reprinted with permission from the Copper Development Association (CDA), www.copper.org/resources/properties/.

2.1.7.16 C83600 Leaded Ounce Metal, or Gunmetal

Alloy C83600 is a copper bearing casting alloy used for numerous poured items, including valves, pump bodies, hardware and ornamental fixtures where a high degree of strength and hardness is not required. Its lead content provides easy machining for secondary operations after being cast.

Table 2.25 Properties of C83600 Leaded Ounce Metal, or Gunmetal*

	Metric	English
Density	8.83 g/cm^3 @ 20°C	0.318 lb/in^3 @ 68°F
Melting Point	1010°C	1850°F
Electrical Conductivity	0.087 MS/cm @ 20°C	15% IACS @ 68°F
Thermal Conductivity		
Hardness (as cast, M01 temper)	Brinell 60	Brinell 60
Yield Strength	117 MPa	17 ksi
Machinability	84%	84%

Chemical Composition (Nominal)

	Metric	English
Copper, Cu	85%	85%

Table 2.25 Properties of C83600 Leaded Ounce Metal, or Gunmetal[*] *(cont.)*

	Metric	English
Lead, Pb	5.0%	5.0%
Tin, Sn	5.0%	5.0%
Zinc, Zn	5.0%	5.0%

[*]Selected data reprinted with permission from the Copper Development Association (CDA), www.copper.org/resources/properties/.

2.1.7.17 C86300 Manganese Bronze

Alloy C86300 is a strong, high yielding casting alloy. Although typically called a bronze, this alloy contains no tin and is technically a brass. Manganese bronzes are able to withstand extreme loads under corrosive conditions, making them useful as ships' propellers and heavily loaded bearings and gears.

Table 2.26   Properties of Manganese Bronze[*]

	Metric	English
Density	F7.83 g/cm^3 @ 20°C	0.283 lb/in^3 @ 68°F
Melting Point	923°C	1693°F
Electrical Conductivity	0.046 MS/cm @ 20°C	8%IACS @ 68°F
Thermal Conductivity	68F35.5 W/m·K @ 20°C	20.50 Btu·ft/(hr·ft^2·°F) @ 68°F
Hardness (Cast, M01 temper)	Brinell 225	Brinell 225
Yield Strength	427 MPa	62 ksi
Machinability	8%	8%

Table 2.26   Properties of Manganese Bronze[*] _(cont.)_

	Metric	English
Chemical Composition		
Copper, Cu	63%	63%
Zinc, Zn	25%	25%
Aluminum, Al	6.2%	6.2%
Iron, Fe	3.0%	3.0%
Manganese, Mn	3.7%	3.7%

[*]Selected data reprinted with permission from the Copper Development Association (CDA), www.copper.org/resources/properties/.

2.1.7.18 C93200 Leaded Bearing Bronze

Alloy C93200 is a commonly used bearing bronze for lathe turned bushings, thrust washers, and plain bearings where metal-to-metal contact occurs. Bronze bearings are sacrificial, being replaced as they wear while protecting the machinery they are installed on. This alloy machines well, has good wear resistance, and offers excellent anti-friction qualities. It resists many chemicals and corrosive atmospheres and is suitable for medium loads and speeds. To save costs on materials and machining time, it is often sold as hollow bar stock in various diameters, ready to size as needed.

Table 2.27   Properties of C93200 Bearing Bronze[*]

	Metric	English
Density	8.91 g/cm^3 @ 20°C	0.322 lb/in^3 @ 68°F
Melting Point	977°C	1790°F

Table 2.27   Properties of C93200 Bearing Bronze [*] *(cont.)*

	Metric	English
Electrical Conductivity	0.07 MS/cm @ 20°C	12% IACS @ 68°F
Thermal Conductivity	58.2 W/m·K @ 20°C	33.60 Btu·ft/(hr·ft^2·°F) @ 68°F
Hardness (M01 temper)	Brinell 65	Brinell 65
Yield Strength	124 MPa	18 ksi
Machinability	70%	70%

Chemical Composition

	Metric	English
Copper, Cu	83%	83%
Lead, Pb	7.0%	7.0%
Tin, Sn	6.9%	6.9%
Zinc, Zn	2.5%	2.5%

[*]Selected data reprinted with permission from the Copper Development Association (CDA), www.copper.org/resources/properties/.

2.1.7.19 *C95400 Aluminum Bronze*

C95400 (Fig. 2.6) is the most commonly used alloy of aluminum bronze, and it finds its uses based on its corrosion resistant and bearing properties. The additional of aluminum forms an oxide coating on the material's exterior, protecting it from corrosion better than straight bronzes. Alloy C95400 is frequently used for bearings and bushings where heavy loads prevail. It is a tough, strong material and is used as a replacement for beryllium copper in non-sparking tools due to its nontoxicity. It also has a biostatic effect when used on ships' hulls and slows the growth of barnacles.

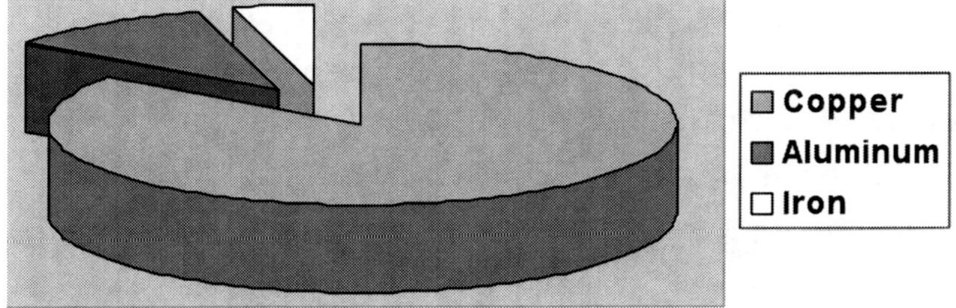

FIGURE 2.6 Elemental makeup of C95400 aluminum bronze.

Table 2.28   Properties of C95400 Aluminum Bronze[*]

	Metric	English
Density	0.269 lb/in^3 @ 68°F	7.45 g/cm^3 @ 20°C
Melting Point	1900°F	1038°C
Electrical Conductivity	13%IACS @ 68°F	0.075 MS/cm @ 20°C
Thermal Conductivity	33.90 Btu·ft/(hr·ft^2·°F) @ 68°F	58.7 W/m·K @ 20°C
Hardness (Cast, M01 temper)	Brinell 170HB	Brinell 170HB
Yield Strength	35ksi	241MPa
Machinability	60%	60%

Chemical Composition (Nominal)

	Metric	English
Copper, Cu	83.2%	83.2%
Aluminum, Al	10.8%	10.8%
Iron, Fe	4.0%	4.0%

[*]Selected data reprinted with permission from the Copper Development Association (CDA), www.copper.org/resources/properties/.

2.2 ALUMINUM AND ITS ALLOYS

2.2.1 Aluminum

Aluminum is a lightweight, silvery-colored metallic element; symbol Al, atomic number 13. In its pure state, aluminum is soft, ductile, and malleable. When exposed to the oxygen in the air, aluminum develops a thin oxide coating that protects the metal and allows it to be used uncoated for many applications. Aluminum can be polished to a highly reflective mirror finish and is an excellent conductor of heat and electricity.

More often than not, aluminum is alloyed with other elements to improve desirable qualities, including hardness and strength. There are two classes of aluminum alloys; wrought and cast.

2.2.1.1 Wrought Aluminum Alloys

There are 490 different alloys of wrought aluminum registered to the Aluminum Association of the United States, and to successfully chart and categorize these alloys, they have created the following system. The designation consists of a four-digit number that, when decoded, designates the major alloy series down to the specific alloy itself. Known as The International Alloy Designation System, it has found worldwide acceptance since being introduced in the 1970s.

Series	Major Alloying Element
1xxx	Straight Aluminum
2xxx	Copper
3xxx	Manganese
4xxx	Silicon

Series	Major Alloying Element
5xxx	Magnesium
6xxx	Magnesium and Silicon
7xxx	Zinc
8xxx	Other

The second digit, if other than 0, represents modifications to the original alloying element in numerical order. For example, a 2411 aluminum has had four modifications from its original 2011 alloy.

In the 1xxx series aluminum, the final two digits represent the minimum purity above 99 percent; i.e., alloy 1045 is 99.45% pure aluminum. In the 2xxx to 8xxx series, the final two digits are simply assigned to designate specific alloys.

2.2.1.2 Cast Aluminum Alloys

Aluminum casting alloys are designated by a four-digit number with a decimal between the third and fourth digits. The first digit specifies the major alloying element or elements.

Series	Major Alloying Element
1xx.x	99.000% Minimum Aluminum
2xx.x	Copper
3xx.x	Silicon Plus Copper and/or Magnesium
4xx.x	Silicon
5xx.x	Magnesium
6xx.x	Unused
7xx.x	Zinc

Series	Major Alloying Element
8xx.x	Tin
9xx.x	Other

The second and third digits signify the specific alloy in its series. The number that follows the decimal indicates the product form. If it is supplied as a casting, that number will be 0, and if supplied as an ingot, it will be 1 or 2.

2.2.1.3 Hardening and Tempering of Aluminum

Aluminum, even when alloyed, is often too soft to be structurally functional or easily machined. Before it leaves the mill, most aluminums are hardened either by heat treatment or strain hardening. Due to the many methods involved in this, a system of coded letters and numbers is used to designate the specific hardening used. There are four major groups, assigned the following letters:

"F"	As fabricated from the mill with no hardening or treatment
"O"	In its dead soft, annealed state
"H"	Strain Hardened (for wrought products only)
"T"	Heat Treated

For strain hardened and heat treated aluminum, there are further codes that specify specific tempering actions taken.

The "H" codes for strain hardened aluminum are:

H1	Strain Hardened only
H2	Strain Hardened and partially annealed

H3	Strain Hardened and stabilized
H4	Strain Hardened and partially annealed by applying a surface coating at high temperatures; seldom used

A second digit also follows, which designates the degree of hardness.

Hx1	1/8 hard	Hx6	3/4 hard
Hx2	1/4 hard	Hx7	7/8 hard
Hx3	3/8 hard	Hx8	full hard
Hx4	1/2 hard	Hx9	extra hard
Hx5	5/8 hard		

A third digit is sometimes used to designate further variations of the "H" tempers. The "T" codes for heat treated aluminum are:

T1	Naturally aged to a stable condition
T2	Annealed (applies to cast products only)
T3	Heat treated, cold worked, and then naturally aged
T4	Heat treated and then naturally aged
T5	Cooled from an elevated temperature and then naturally aged
T6	Heat treated and then artificially aged
T7	Heat treated and then stabilized
T8	Heat treated, cold worked, and then artificially aged
T9	Heat treated, artificially aged, and then cold worked (note that this is resequenced from T8)
T10	Artificially aged and then cold worked

One or two additional digits are sometimes applied that designate additional stress relieving by various methods.

2.2.1.4 1100 Aluminum

Alloy 1100 aluminum, known as "commercially pure," has a minimum purity of 99.0% aluminum. Being basically unalloyed, it is relatively soft and easily formed. It is the easiest aluminum alloy to weld, the most resistant to corrosion, and the most conductive of heat and electricity. Alloy 1100 cannot be heat treated but can be successfully cold worked to a high (H18) temper. Due to the gumminess of alloy 1100, it can only be machined at this high hardness. Uses of this alloy include foil, rivets, and tie wire that take advantage of its formability. It is a relatively weak aluminum and lacks the strength of most other aluminum alloys.

Table 2.29   Properties of H12 Temper[*]

Chemical Composition	
Manganese	0.050%
Magnesium	—
Zinc	0.10%
Titanium	—
Vanadium	—
Iron	—
Silicon	—
Copper	0.050–0.20%
Beryllium	0.000800%
Silicon/Iron	0.95%
Others, total	0.15%
Aluminum	Balance

Table 2.29   Properties of H12 Temper[*] *(cont.)*

Density		
Designation	lb/in^3	g/cm^3
1100	0.098	2.71

[*]Selected data from the "Teal sheets" and from "2009 Aluminum Standards & Data," reprinted with permission from The Aluminum Association Inc., www.aluminum.org.

2.2.1.5 2011 Aluminum

This is an age hardenable, wrought alloy noted for its free machining qualities. The addition of lead and bismuth (approximately 0.5 percent each) allows ease of machining, super finishes, and long tool life, making this alloy a favorite in the machine shop. Alloy 2011 is heat treatable. Its main alloying metal is copper (5 to 6 percent), which gives this alloy high strength but hinders weldability.

Table 2.30   Properties of 2011-T6[*]

Chemical Composition	
Manganese	—
Magnesium	—
Zinc	—
Titanium	—
Vanadium	—
Silicon	—
Copper	5.0–6.0%
Beryllium	—
Silicon/Iron	—

Table 2.30   Properties of 2011-T6[*] *(cont.)*

Bismuth	0.20–0.60%
Iron	0.70%
Lead	0.20–0.60%
Other, Total	0.15%
Aluminum	Balance

Density

Designation	lb/in^3	g/cm^3
2011	0.102	2.823

[*]Selected data from the "Teal sheets" and from "2009 Aluminum Standards & Data," reprinted with permission from The Aluminum Association Inc., www.aluminum.org.

2.2.1.6 2024 Aluminum

Alloy 2024 aluminum has the highest tensile strength of any of the common aluminum alloys, approaching and even exceeding that of mild steel. It is easily machined, making it a favorite material for high strength screw machine products including bolts, shafts, and similar items. Other uses include aircraft frame and fuselage members that take advantage of the material's strength and rigidly. When compared to 6061, however, it is not nearly as weldable or resistant to corrosion.

Table 2.31   Properties of 2024-T6[*]

Chemical Composition	
Manganese	0.30–0.90%
Magnesium	1.20–1.80%
Zinc	—

Table 2.31   Properties of 2024-T6[*] *(cont.)*

Titanium	—
Vanadium	—
Iron	0.50%
Silicon	—
Copper	—
Beryllium	—
Silicon/Iron	—
Chromium	0.10%
Other, Total	0.15%
Aluminum	Balance

Density

Designation	lb/in^3	g/cm^3
2024	0.100	2.768

[*]Selected data from the "Teal sheets" and from "2009 Aluminum Standards & Data," reprinted with permission from The Aluminum Association Inc., www.aluminum.org.

2.2.1.7 5052 Aluminum

Alloy 5052 is typically used in sheet and tube forms. It is readily weldable and has a very high resistance to corrosion, including in marine environments. Its main alloying element is magnesium (typically 2.5 percent). Alloy 5052 is easily cold worked and has very high fatigue strength. These properties make it useful in the transportation field for boat hulls, fuel lines, and other components where frequent stresses and vibrations are found. Its machinability rating is only fair to good, and it is not generally used as lathe or milling stock. Alloy 5052 is not heat treatable but can be hardened by cold working.

Table 2.32   Properties of 5052-H34 Temper[*]

Chemical Composition

Manganese	0.10%
Magnesium	2.20–2.80%
Zinc	—
Titanium	—
Vanadium	—
Iron	0.40%
Silicon	0.25%
Copper	0.10%
Beryllium	—
Silicon/Iron	—
Other, total	0.15%
Aluminum	Balance

Density

Designation	lb/in^3	g/cm^3
5052	0.097	2.685

[*]Selected data from the "Teal sheets" and from "2009 Aluminum Standards & Data," reprinted with permission from The Aluminum Association Inc., www.aluminum.org.

2.2.1.8 6061 Aluminum

Alloy 6061 is a general-purpose aluminum alloy featuring fair machinability and excellent corrosion resistance. It is easily welded, offers high strength and good resistance to stress corrosion cracking. Alloy 6061 is widely available and relatively inexpensive, making this metal very popular for items ranging from aeronautical couplings and fixtures to items in the hobby shop.

Alloy 6061 is heat treatable, weldable, and easily cold and hot worked. When lathe turning 6061, it is advisable to use a chip-breaker tool to prevent the formation of long, stringy turnings that can be a nuisance to the operator. If cutting tools are kept sharp, excellent finishes can be produced on this and most aluminum alloys. A good flood or mist of coolant is necessary, as even the hardest aluminum alloys will gum up on the cutting tool otherwise (Fig. 2.7).

Table 2.33   Properties of 6061-T6, T651[*]

Chemical Composition

Manganese	0.15%
Magnesium	0.80–1.20%
Zinc	0.25%
Titanium	0.15%
Vanadium	—
Iron	0.70%
Silicon	0.40–0.80%
Copper	0.15–0.40%
Beryllium	—
Silicon/Iron	—
Other, total	0.15%
Aluminum	Balance

Density

Designation	lb/in^3	g/cm^3
6061	0.098	2.713

[*]Selected data from the "Teal sheets" and from "2009 Aluminum Standards & Data," reprinted with permission from The Aluminum Association Inc., www.aluminum.org.

FIGURE 2.7 Aluminum round rod and tubing in alloy 6061, temper T6.

2.2.1.9 6063 Aluminum

Alloy 6063 aluminum is frequently used as architectural items due to its ease in formability, excellent corrosion resistance, weldability and high surface finish. Its main alloying element is magnesium (typically 0.7%). Uses of alloy 6063 include window and door frames, railing, piping and other extrusions. It also has very good thermal conductivity, making it useful for heat sinks. Alloy 6063 is heat treatable, and when hardened it can be machined with good results.

Table 2.34   Properties of 6063-T6 Temper[*]

Chemical Composition	
Manganese	0.10%
Magnesium	0.45–0.90%
Zinc	0.10%
Titanium	0.10%
Vanadium	—
Iron	0.35%
Silicon	0.20–0.60%
Copper	0.10%
Beryllium	—
Silicon/Iron	—
Other, total	0.15%
Aluminum	Balance

Density		
Designation	lb/in^3	g/cm^3
6063	0.097	2.685

[*]Selected data from the "Teal sheets" and from "2009 Aluminum Standards & Data," reprinted with permission from The Aluminum Association Inc., www.aluminum.org.

2.2.1.10 7075 Aluminum

When the highest strength-to-weight ratio of any aluminum is needed, alloy 7075 is used. With a typical tensile strength of 83 ksi, it is stronger than 1018 steel by several percentage points while having less than half of its density. It has good machinability, fair corrosion resistance, and poor weldability relative to other aluminum alloys. Its main alloying element is zinc (typically 5.5 percent). Due to its superb strength, it is used for high-stress parts, including aircraft landing gear, struts, and fuselage components. Other uses include screw machine products where high strength and good machinability are needed.

Table 2.35   Properties of 7075-T6, T651[*]

Chemical Composition	
Manganese	0.30%
Magnesium	2.10–2.90%
Zinc	5.10–6.10%
Titanium	0.20%
Vanadium	—
Iron	0.50%
Silicon	0.40%
Copper	1.20–2.0%
Beryllium	—
Silicon/Iron	—
Other, total	0.15%
Aluminum	Balance

Density		
Designation	lb/in^3	g/cm^3
7075	0.101	2.796

[*]Selected data from the "Teal sheets" and from "2009 Aluminum Standards & Data," reprinted with permission from The Aluminum Association Inc., www.aluminum.org.

2.3 NICKEL AND ITS ALLOYS

2.3.1 Nickel

Nickel is a hard, silvery white metallic element, symbol Ni, atomic number 28. Pure nickel is not found in nature; rather, it is extracted from ores. Primary ores include nickel/iron pyrites and nickel arsenides. Nickel is malleable, ductile, and one of the few elements that are magnetic. It is a good conductor of heat and electricity, can take on a high polish, and is very resistant to corrosion. Nickel is difficult to machine due to the material's toughness and gumminess. It can be accomplished, however, with very sharp tools having positive rake angles and rigid set-ups. It can be readily soldered with proper fluxes and can also be easily welded.

Pure nickel has limited uses, but its alloys are numerous and widely used. Pure nickel is used as a plating metal, in recharge-able batteries, and as a chemical catalyst. Much more prevalent, however, are the hundreds of commercial alloys that contain nickel. These include the stainless steels, cupronickel, Monel, and the nickel superalloys.

Table 2.36   Properties of Nickel[*]

	Metric	English
Density	8.88 g/cm^3	0.321 lb/in^3
Tensile Strength, Ultimate	45.0 MPa	6530 psi
Tensile Strength, Yield	59.0 MPa	8560 psi
Elongation at Break	30.0%	30.0%
Melting Point	1455°C	2651°F

Table 2.36   Properties of Nickel [*] ***(cont.)***

	Metric	English
Chemical Composition		
Nickel	100%	100%

[*]Selected data from the MatWeb search database reprinted with permission from MatWeb, www.matweb.com.

2.3.2 Monel

Monel is a trademarked name for high-performance nickel alloys that are strong, corrosion resistant, and chemical resistant. Alloy 400 (UNS N04400) is the most common Monel alloy used, and it has excellent ductility and weldability. It can be used in temperatures ranging from sub-zero to 1000°F and has been in use for over 100 years. Alloy 500 (UNS N05500) has added aluminum and titanium and is totally nonmagnetic. It has twice the tensile strength of alloy 400 and is precipitation heat treatable for further strength.

Typical uses for these alloys include propeller shafts, pumps, valves, and general hardware for use in marine or otherwise corrosive environments. Nonmagnetic alloy 500 is used for gyroscopes and compass enclosures on marine vessels. Monel is difficult to machine and work hardens easily.

Table 2.37   Properties of Monel [*]

	Metric	English
Density	8.80 g/cm^3	0.318 lb/in^3
Tensile Strength, Ultimate	550 MPa	79,800 psi

Table 2.37 Properties of Monel[*] *(cont.)*

	Metric	English
Tensile Strength, Yield	240 MPa	34,800 psi
Elongation at Break	48.0%	48.0%
Melting Point	1300–1350°C	2370–2460°F

Chemical Composition

	Metric	English
Carbon, C	≤0.30%	≤0.30%
Copper, Cu	28.0–34.0%	28.0–34.0%
Iron, Fe	≤2.50%	≤2.50%
Manganese, Mn	≤2.0%	≤2.0%
Nickel, Ni	≥63.0%	≥63.0%
Silicon, Si	≤0.50%	≤0.50%
Sulfur, S	≤0.0240%	≤0.0240%

[*]Selected data from the MatWeb search database reprinted with permission from MatWeb, www.matweb.com.

2.3.2.1 Kovar (UNS K94610)

Kovar (Fig. 2.8) is a nickel iron cobalt alloy designed to have nearly the exact same thermal expansion as borosilicate glass. Keeping this expansion identical allows Kovar to be used as hermetic seals along with the glass in items ranging from microwave and X-ray tubes to semiconductor parts in the electronics field. Kovar has good mechanical properties and can be hot and cold worked. Like other high nickel alloys, it is difficult to machine due to its gumminess and toughness, but it can be accomplished with proper feeds, speeds, and tool geometries.

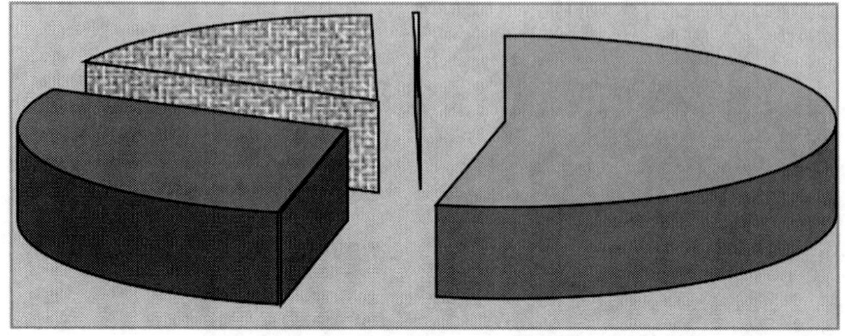

FIGURE 2.8 A look at Kovar's makeup.

Table 2.38   Properties of Kovar[*]

	Metric	English
Density	8.36 g/cm^3	0.302 lb/in^3
Tensile Strength, Ultimate	517 MPa	75,000 psi
Tensile Strength, Yield	345 MPa	50,000 psi
Elongation at Break (%)	30.0%	30.0%

Chemical Composition

	Metric	English
Carbon, C	≤0.020%	≤0.020%
Cobalt, Co	17.0%	17.0%
Iron, Fe	53.0%	53.0%
Manganese, Mn	0.30%	0.30%
Nickel, Ni	29.0%	29.0%
Silicon, Si	0.20%	0.20%

[*]Selected data from the MatWeb search database reprinted with permission from MatWeb, www.matweb.com.

2.3.2.2 Inconel 600 (UNS N06600), N06601, N07718)

Inconel is a nickel/chrome/iron alloy noted for its excellent resistance to heat and corrosion. As one of the nickel "superalloys," it is widely used in the aeronautical field in engine components and exhaust liners. It maintains its properties at cryogenic temperatures without becoming brittle, and can withstand heat of up to 2000°F without failure.

Inconel has good strength and is easily hot and cold worked, but like many nickel alloys, it is difficult to machine. Different Inconel alloys are produced having special characteristics. Alloy 601 (UNS N06601) contains aluminum that leads to better mechanical properties at elevated temperatures over alloy 600. Alloy 718 (UNS N07718) is heat treatable and can be easily welded.

Table 2.39   Properties of Inconel 600[*]

	Metric	English
Density	8.47 g/cm^3	0.306 lb/in^3
Tensile Strength, Ultimate	655 MPa	95,000 psi
Tensile Strength, Yield	310 MPa	45,000 psi
Elongation at Break	45.0%	45.0%
Melting Point	1354–1413°C	2469–2575°F

Chemical Composition

	Metric	English
Carbon, C	≤0.15%	≤0.15%
Chromium, Cr	14.0–17.0%	14.0–17.0%
Copper, Cu	≤0.50%	≤0.50%
Iron, Fe	6.0–10.0%	6.0–10.0%

Table 2.39   Properties of Inconel 600[*] *(cont.)*

	Metric	English
Manganese, Mn	≤1.0%	≤1.0%
Nickel, Ni	≥72.0%	≥72.0%
Silicon, Si	≤0.50%	≤0.50%
Sulfur, S	≤0.015%	≤0.015%

[*]Selected data from the MatWeb search database reprinted with permission from MatWeb, www.matweb.com.

2.3.2.3 *Multimet N155 (UNS R30155)*

Multimet is a high chrome, cobalt, nickel, and molybdenum alloy used for its high-temperature qualities. It was one of the early nickel superalloys produced, dating back to the late 1940s. Multimet has excellent corrosion and oxidation resistance while maintaining its strength and hardness in a continuous temperature of 1500°F. It is an easily welded material and can be hot forged and machined with good results. While cold working, it is easily work hardened, and annealing is often required between steps. Multimet is heat treatable and available in extruded shapes as well as castings. Typical uses include turbine blades, exhaust liners, afterburner parts, and high-temp hardware.

Table 2.40   Properties of Multimet[*]

	Metric	English
Density	8.20 g/cm^3	0.296 lb/in^3
Tensile Strength, Ultimate	800 MPa	116,000 psi
Tensile Strength, Yield	393 MPa	57,000 psi

Table 2.40   Properties of Multimet[*] *(cont.)*

	Metric	English
Elongation at Break (%)	43.0%	43.0%
Melting Point	1288–1354°C	2350–2469°F

Chemical Composition

	Metric	English
Carbon, C	0.080–0.16%	0.080–0.16%
Cb + Ta	0.75–1.25%	0.75–1.25%
Chromium, Cr	20.0–22.5%	20.0–22.5%
Cobalt, Co	18.5–21.0%	18.5–21.0%
Iron, Fe	33.0%	33.0%
Manganese, Mn	1.0–2.0%	1.0–2.0%
Molybdenum, Mo	2.50–3.50%	2.50–3.50%
Nickel, Ni	19.0–21.0%	19.0–21.0%
Nitrogen, N	0.10–0.20%	0.10–0.20%
Silicon, Si	≤1.0%	≤1.0%
Tungsten, W	2.0–3.0%	2.0–3.0%

[*]Selected data from the MatWeb search database reprinted with permission from MatWeb, www.matweb.com.

2.3.2.4 Invar (UNS K93600 and K93650)

Invar is a nickel iron alloy featuring extremely low thermal expansion. This property makes it useful in precision clocks, instruments, and very accurate tape measures. Invar is also used with a metal of higher thermal expansion in the makeup of bimetallic strips that are used in thermostats and thermometers. When these two metals are coupled together, the metal with the higher expansion will curve outward when heated, and inward when cooled.

Invar is very difficult to machine, being tough, gummy, and readily work hardened. A free machining variety (UNS K93650) has been made that contains a small amount of selenium, which makes machining tolerable.

Table 2.41   Properties of Invar[*]

	Metric	English
Density	8.05 g/cm^3	0.291 lb/in^3
Tensile Strength, Ultimate	462 MPa	67,000 psi
Tensile Strength, Yield	261 MPa	37,900 psi
Elongation at Break (%)	32.0%	32.0%
Chemical Composition		
Aluminum, Al	≤0.030%	≤0.030%
Carbon, C	0.0080%	0.0080%
Cobalt, Co	≤0.35%	≤0.35%
Iron, Fe	63.0%	63.0%
Manganese, Mn	0.30%	0.30%
Molybdenum, Mo	≤0.020%	≤0.020%
Nickel, Ni	36.0%	36.0%
Silicon, Si	0.15%	0.15%
Sulfur, S	0.0010%	0.0010%

[*]Selected data from the MatWeb search database reprinted with permission from MatWeb, www.matweb.com.

2.3.2.5 Mu-Metal

Mu-metal and similar nickel iron alloys have the highest magnetic permeability of any materials (some 10,000 times that of steel) and are used as magnetic shielding. Mu-metal works by absorbing and dissipating the magnetic field, keeping it away from what it is protecting. It is known as a "soft" magnetic material, losing its magnetism when the source is removed.

Mu-metal is found in hospital MRI rooms, computer equipment, and other electronics. In laboratory work, it is used to isolate sensitive experiments and provide a space relatively free from magnetic flux. Mu-metal is available in sheet form and can be cold worked with common methods. Once in its final shape, mu-metal must be heat treated: this treatment increases its permeability by several times. Working on mu-metal after heat treatment misaligns its crystalline structure and lowers its permeability.

2.3.2.6 Nichrome

Nichrome is a nickel and chrome alloy commonly drawn into wire and ribbon for use in heating elements. Nichrome is resistant to corrosion, has a high melting point, and has a very high electrical resistance—three properties needed to make a proper heating element material. All electrical resistance heating, including water heaters and home heating, toasters, hair dryers, and ovens use nichrome to transfer electricity into heat. Different grades of nichrome are made having specific resistive properties as needed. It is gray in color.

Table 2.42 Properties of Nichrome[*]

	Metric	English
Density	7.90 g/cm^3	0.285 lb/in^3
Tensile Strength, Ultimate	750 MPa	109,000 psi

Table 2.42   Properties of Nichrome[*] *(cont.)*

	Metric	English
Tensile Strength, Yield	450 MPa	65,300 psi
Elongation at Break (%)	30.0%	30.0%
Melting Point	1390°C	2530°F

Chemical Composition

	Metric	English
Chromium, Cr	20.0%	20.0%
Iron, Fe	42.0%	42.0%
Manganese, Mn	1.00%	1.00%
Nickel, Ni	35.0%	35.0%
Silicon, Si	2.00%	2.00%

[*]Selected data from the MatWeb search database reprinted with permission from MatWeb, www.matweb.com.

2.3.2.7 Nitinol

The name Nitinol is an acronym for *nickel titanium naval ordnance laboratory*. It is a nickel titanium alloy that has some incredible properties, including shape memory and super elasticity. When a shape memory Nitinol alloy is deformed from its original shape, it will snap back into place with great force when heated above its transformation temperature. These forces can be harnessed to produce push and pull movements in mechanisms, provided that the required temperatures are present.

Certain Nitinol alloys also possess super-elasticity properties. These alloys, when deformed, will spring back to original as soon as the load is removed, with no heat needed. Their elastic properties exceed spring steel's by a factor of 20. Super-elastic

Nitinol alloys are used for many items, including eyeglass frames and medical stents.

Table 2.43 Properties of Nitinol[*]

	Metric	English
Density	6.45 g/cm^3	0.233 lb/in^3
Tensile Strength, Ultimate	754–960 MPa	109,000–139,000 psi
Tensile Strength, Yield	560 MPa	81,200 psi
Elongation at Break (%)	15.5%	15.5%
Melting Point	1240–1310°C	2260–2390°F

Chemical Composition

	Metric	English
Nickel, Ni	55.0%	55.0%
Titanium, Ti	45.0%	45.0%

[*]Selected data from the MatWeb search database reprinted with permission from MatWeb, www.matweb.com.

2.4 LOW-MELTING METALS

2.4.1 Lead

Lead (atomic number 82, symbol Pb) is a heavy, soft metallic element. It is highly malleable, ductile, and corrosion resistant. Its color is a bright silvery blue, but it rapidly tarnishes to a dull gray when exposed to the oxygen in the air. It has a very low melting point and, relative to other metals, it is a poor conductor of electricity. Lead has been shown to be poisonous, and it accumulates in the body over time. It is a common, inexpensive metal used worldwide for many purposes.

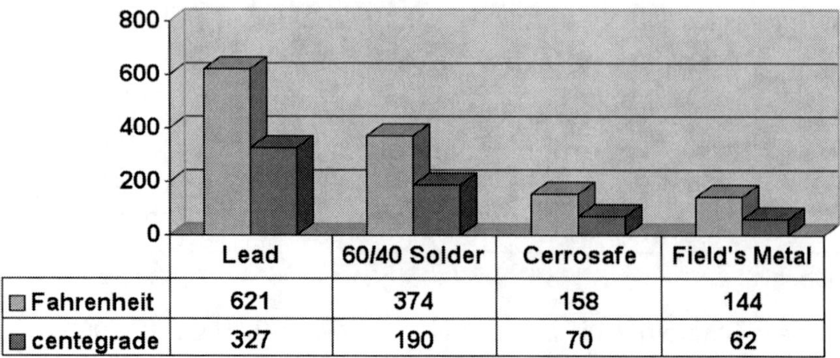

FIGURE 2.9 Melting temperatures of low-melt metals.

Like all materials, lead finds its uses based on its properties. Its high density makes it useful as ammunition projectiles, fishing sinkers, wheel balance weights, and radiation shields. Its corrosion resistant properties enable it to be used in sheeting for lining chemical tanks. Similar sheeting is also used in the building trades for sound and vibration damping. Lead's low melting temperature makes it useful as a fusible metal in fire sprinklers. Its main use, however, is for grid plates that made up lead acid storage batteries.

Lead is often alloyed with other metals to improve its properties. Adding small amounts of cadmium, copper, or antimony increases leads hardness and strength considerably, while lead and tin mixtures are used as solders, organ pipes, and babbitt bearings.

Table 2.44   Properties of Lead[*]

	Metric	English
Density	10.22 g/cm^3 @ 800°C	0.3692 lb/in^3 @ 1470°F
Hardness, Brinell	4.2	4.2

Table 2.44   Properties of Lead[*] *(cont.)*

	Metric	English
Melting Point	327.5°C	621.5°F
Tensile Strength, Ultimate	18.0 MPa	2610 psi
Yield Strength, Ultimate	8 MPa	1160 psi

[*]Selected data from the MatWeb search database reprinted with permission from MatWeb, www.matweb.com.

2.4.2 Babbitt

Babbitt refers to various alloys of tin- or lead-based materials. These alloys are used as plain bearings on rotating shafts in machinery. Babbitt bearings are usually poured into place after aligning the shaft inside of a pillow block housing. Although roller and needle bearings have replaced Babbitt on many items, it still can frequently be found on larger equipment.

Two types of Babbitt metal are common; tin-based and lead-based. Tin-based Babbitt can typically run twice as fast and carry four times the load of lead-based. The original formula created by Isaac Babbitt in 1839 was composed of 89.3 percent tin, 7.1 percent antimony, and 3.6 percent copper. It is now known as "type 2 original Babbitt," UNS-L13890. Lead-based Babbitt is less expensive and adequate for slower speeds and lighter loads.

Table 2.45   Typical Properties of Lead-Based Babbitt[*]

	Metric	English
Density	10.6 g/cm^3	0.383 lb/in^3
Hardness, Brinell	19	19

Table 2.45   Typical Properties of Lead-Based Babbitt[*]

	Metric	English
Melting Point	240–268°C	464–514°F
Tensile Strength, Ultimate	70.0 MPa	10,200 psi
Yield Strength, Ultimate	10 MPa	1450 psi

[*]Selected data from the MatWeb search database reprinted with permission from MatWeb, www.matweb.com.

2.5 FUSIBLE ALLOYS

Fusible alloys are a family of metals with low melting temperatures relative to other metals. Some of these alloys are liquid at below room temperature and have a very interesting property—that of having a much lower melting temperature of any of its alloying elements. Fusible alloys are dense, silvery metals made from combinations of bismuth, antimony, lead, gallium, cadmium, zinc, and indium. Lead and cadmium are hazardous to the body, and precautions need to be taken when working with these metals. Woods metal, e.g., has a melting temperature of 158°F (70°C).

2.5.1 Field's Metal

Field's metal is a nontoxic replacement for Wood's metal, having neither lead nor cadmium in its makeup. It has a eutectic melting/solidifying temperature of 144°F and has a formula of 32.5 percent Bi, 51 percent In, and 16.5 percent Sn.

2.5.2 Cerrosafe

Cerrosafe is a non-eutectic formula used by gunsmiths to cast and measure cartridge chambers. It has the unique property among fusible alloys in that it will temporarily shrink in size for the first 30 minutes of cooling, allowing it to be easily removed from the chamber. In one hour's time after pouring, it will have expanded to exact chamber size, allowing precision measurements to be made. It has a melting/solidifying range of 158°F to 190°F. Its formula is 49 percent Bi, 37.7 percent Pb, 11.3 percent Sn, and 8.5 percent Ca.

2.5.3 Galinstan

Galinstan is a eutectic blend of 68.5 percent Ga, 21.5 percent In, and 10 percent Sn, having a melting/solidifying temperature of −2°F. It is nontoxic and is a replacement for mercury in thermometers, thermostats, and tilt switches. Unlike mercury, however, it tends to spread out and wet surfaces. To prevent this, gallium oxide coatings are used in the above instruments to keep the metal afloat. Its wetting properties can be used, however, to coat glass surfaces for use as mirrors.

2.5.4 Pewter

Pewter is a tin alloy with added copper, antimony, and/or lead. Traditionally, pewter has been used for dinnerware, kitchenware, and flatware. Modern pewter is used mainly as decorative figures, vessels, and medallions. Although it is produced with a mostly tin alloy, pewter holds an affluent connotation among its admirers due to its documented history and rich luster when polished. A modern pewter casting alloy (UNS L13911) contains 90–93 percent Sn, 6–8 percent Sb and 0.25–2 percent

Cu. Traditional pewter was a tin/lead alloy, but in modern times it has been made lead-free due to health concerns. The lead-free varieties look nearly identical but lack the blue hue of leaded pewter. Pewter is known (by the pewter manufacturers) as the forth precious metal, after gold, silver, and platinum. Unlike silver, pewter will not tarnish in reaction to air and will hold its appearance for many years.

Table 2.46   Properties of Pewter[*]

	Metric	English
Density	7.28 g/cm^3	0.263 lb/in^3
Hardness, Brinell	8	8
Melting Point	244–295°C	471–563°F
Tensile Strength, Ultimate	52.0 MPa	7540 psi
Ultimate Yield Strength	n/a	n/a

[*]Selected data from the MatWeb search database reprinted with permission from MatWeb, www.matweb.com.

2.5.5 Solder

Solder is a general term for low-melting alloys used to fuse one metal to another. It can be divided into two categories: soft solder and hard solder.

2.5.5.1 Soft Solders

Soft solders have a relatively low melting temperature and are either leaded or lead free. They are available in wire, ribbon, bar, and paste forms. Wire solders are offered either solid or with flux cores.

2.5.5.2 Leaded Solders

Leaded solders are the most common types and are superior to lead-free in performance. Most are alloys of tin/lead, and the balance between the two defines the melting and freezing temperatures of the alloys. When the alloy contains 63 percent tin and 37 percent lead (called 63/37), it has the lowest melting temperature of all tin lead solders (361°F). The 63/37 alloy also is remarkable in that its melting point is the same as its freezing point, with no pasty range as in other tin-lead mixtures. This is called its *eutectic* point, meaning "easily melted" or "all melted at once." This type of solder will melt, wet, and solidify at a much faster pace than other alloys, making quicker, cleaner work while using the lowest possible soldering temperature.

The most popular solder for general-purpose work is 60/40. It has a 13°F paste range (melts at 374°F and hardens at 361°F) and leaves a nice shiny finish. The cost for 60/40 is considerably less than that of 63/37 and is a good choice for most electronics and general work.

Alloy 50/50 solder melts at 421°F and hardens at 361°F, having a paste range of 60°F. This property is useful in situations where the solder has to be manipulated into place or wiped away with a towel before it hardens. 50/50 solder also leaves a shorter bead as compared to higher-tin solders.

2.5.5.3 Lead-Free Solders

Due to concerns over lead in drinking water systems and landfills, lead-free alloys have been produced. Although their performance is substandard to that of leaded solders, good joints can be achieved with the proper techniques and fluxes. Lead-free solders generally have higher melting temperatures, inferior physical properties, and more difficult application than leaded varieties. Typical makeup for lead-free solders includes tin, silver, and copper with antimony and bismuth also added.

For the purpose of soldering aluminum, different tin/zinc alloys have been produced, including 91/9 and 60/40. Due to the heavy oxide layer that persists on aluminum, heavy-duty fluxes are needed to penetrate this layer and wet the aluminum.

2.5.5.4 Hard Solder (Silver Solder)

Hard solders are often silver bearing with added copper and zinc. Having a much higher melting temperature than soft solders, they are used in making jewelry and for high-stress or high-pressure connections. Hard solders are also used where higher temperatures will be encountered, including cookware and teapots. Gold-bearing solders are available for making jewelry in different colors in order to match specific carat purities. Their composition is varying amounts of gold, copper, and silver with zinc added to lower the melting temperature.

2.5.6 Type Metal

Type metals are lead alloys used in old style metal type characters. To strengthen the lead, antimony and tin are added, which also makes an alloy that slightly expands upon solidifying; this gives the type very crisp lines as it is taken from its mold. Different alloys are used depending on the typesetting technique. A typical alloy contains 78 percent Pb, 15 percent

Sb, and 7 percent Sn (UNS L53570), although there are literally scores of type metal alloys. Type metals have also been used as general-purpose casting alloys for small figurines and trinkets but have since been replaced with lead-free metals. Today, black powder enthusiasts who cast their own bullets enjoy type metal for its high degree of hardness over pure lead. It is a cheap alternative to engineered bullet alloys and very comparable in properties.

2.6 MAGNETIC MATERIALS

2.6.1 Ferrite (Ceramic) Magnets

Ferrite magnets start off with either strontium ferrite or barium ferrite mixed with iron oxide. This mixture of powders is compressed in a die and then sintered at high temperatures. Ferrite magnets are popular due to their low cost, wide range of available sizes, and high resistances to corrosion and demagnetization. Like most magnet materials, they are very hard and brittle, and they cannot be machined other than with abrasive wheels. Ferritic magnets are made in two types: anisotropic (oriented) and isotropic (non-oriented). In an anisotropic magnet, an oriented magnetic field is applied during the manufacturing process, which aligns the individual grains within the magnet before it solidifies. Afterward, they are again magnetized, making anisotropic magnets the strongest variety of ferritic magnets available. Isotropic magnets, on the other hand, have internal grains and crystals misaligned from each other and have much lower magnetic capacities.

2.6.2 Alnico

Alnico is an acronym for alloys of *aluminum, nickel, and cobalt* that are used in permanent magnets. Alnico magnets have a much higher energy (magnetism) than conventional ceramic magnets and place second behind the rare earth magnets. They have the largest usable temperature range of any of the permanent magnets and can withstand temperatures of up to 1000°F. Alnico magnets are produced by either casting or sintering, depending on the final shape needed. Alnico is a very hard, brittle material that does not machine with conventional methods but must be cut with abrasive wheels. One downfall of this material is its low magnetic coercive force, which is easily weakened and demagnetized.

2.6.3 Neodymium Iron Boron ($Nd_2Fe_{14}B$)

Formulated in 1982, this alloy produces the strongest permanent magnets known. Colloquially known as *rare earth, neo,* or *NIB* magnets, they have not only replaced lesser magnets in many applications, they also have allowed new items to be made that were simply waiting for strong enough magnets to exist.

Neodymium magnets, besides being the strongest magnets for their size, are also the most difficult to demagnetize. They are not rated for high temperatures, though, and regular neodymium grades have a maximum operating temperature of 176°F (80°C) and a permanent failure (Curie) temperature of 590°F (310°C). For high-temperature use, Alnico magnets can be used, which have a maximum operating temperature of 1004°F (540°C) and a Curie temperature of 1580°F (860°C).

Neodymium magnets are extremely hard, brittle, and corrosion susceptible. They cannot be conventionally machined and, like other magnets, are usually initially formed to the exact size needed. Due to their vulnerability to corrosion, they are usually plated at the factory.

2.6.4 Samarium-Cobalt Magnets

The Samarian-cobalt magnet, along with the neodymium magnet, belongs to the rare earth family of magnets due to its uncommon elemental makeup. As with their neodymium cousins, they are strong and brittle, but samarium cobalt magnets have higher operating temperatures (482°F, 250°C) and Curie temperature (1382°F, 750°C). They are not as magnetically strong, however, with the basic grade (grade 18) samarium cobalt magnet being only 2/3 the strength of a basic (grade 27) neodymium magnet. Samarium cobalt magnets have excellent corrosion resistant properties in moist, acid, and alkali environments and do not require plating as do neodymium magnets. They are, however, considerably more expensive than neodymium.

2.7 MISCELLANEOUS NONFERROUS METALS

2.7.1 Chromium

Chromium is a hard, lustrous metallic element, symbol Cr and atomic number 24. It is brittle and silver blue in color, and it takes a high mirror polish that will not tarnish in air. Chromium is not found as a native metal but in chromite (iron magnesium chromium oxide), its only ore.

Chromium is used mainly as an alloying ingredient in stainless steel, where it is coupled with nickel in varying amounts and added to iron to make the different stainless alloys. Its other main use is a plated finish on steel, where its high gloss and corrosion resistant properties appeal to the eye while protecting the plated material from the elements. Chromium plating is also used to increase the material's hardness on bearing surfaces. A thicker application is used for this hard coating than for ornamental use, and its surface can be as hard as 68 on the Rockwell "C" scale. Hard coat chrome is used to coat many items including piston rings, cylinder walls, and camshafts where surfaces bear against others.

2.7.2 Cobalt

Cobalt is a brittle, hard, gray metallic element, symbol Co, atomic number 27. Metallic cobalt is not found in its native state but instead is refined from several ores, namely cobaltite, smaltite, and erythrite. Cobalt, along with iron and nickel, is a ferromagnetic material. These three elements are frequently found in meteorites in their metallic form, producing a hard, rust-resistant curiosity.

Cobalt is seldom used in its pure form, with the exception being an electroplating material. More often, cobalt is used as an alloying component in various steel and nickel compounds called *superalloys*. These high-tech metals display excellent strength at high temperatures, enabling them to be used for jet engine components, for example. Cobalt is also used to strengthen high speed steel cutting tools in the machine shop and as a matrix binder for special types of cobalt magnets.

2.7.3 Magnesium

Magnesium is a silvery white, lightweight metallic element, symbol Mg, atomic number 12. Like aluminum, it develops a corrosion-resistant film on its exterior that protects it from further decay. Magnesium is the lightest engineering metal material, with a weight two thirds that of aluminum and one quarter that of steel. It is the easiest metal to machine, even in its strongest alloys. Magnesium alloys are available in ingots for casting media as well as in wrought form. Finely divided magnesium will burn with a hot, bright white flame, and care must be taken not to overheat it when machining. This exothermic property makes magnesium useful in fireworks and emergency fire starters.

Despite the pyrotechnic properties of magnesium, it is mainly used as an engineering material that benefits from its light weight and high toughness. It is often used for automotive and aerospace components as well as sporting goods. As an alloy, magnesium will increase the hardness and strength of aluminum without adding to its specific gravity. Magnesium is also vital to the human body, which may contain over 30 g of elemental magnesium in a full-grown adult.

2.7.4 Zinc

Zinc is a silvery gray metallic element, symbol Zn, and atomic number 30. At room temperature, zinc is very brittle but, when heated to around 200°F, it becomes soft, malleable, and ductile, enabling it to be worked into sheets or drawn into wires.

Zinc finds its main use as a coating on steel or iron, in a process known as *hot dipped galvanization.* As the name suggests, the objects are passed through molten zinc, which leaves a weatherproof coat that protects against corrosion. A galvanized coating can last from 30 to 100 years, depending on the thickness of the application.

Zinc, when alloyed with copper, produces brass. This alloy has been known since antiquity, and brass was produced from various ores before zinc itself was even isolated.

Today, zinc is used as a die casting metal and is often alloyed with other low-temperature melting metals to produce desired properties. Due to the rising cost of copper, U.S. pennies minted since 1983 have a zinc content of 97.5 percent.

2.7.5 Tin

Tin is a soft, silvery, malleable, and ductile metallic element (symbol Sn, atomic number 50) used mainly as an alloying ingredient in solder, pewter, and bronze. Being resistant to corrosion, it is plated on steel to store food items in "tin cans" and also on architectural roofing and other steel exterior panels.

When pure tin is cooled below 56°F (13.2°C) for long periods, it deteriorates into a gray powder, losing all of its metallic properties. Known as *tin pest,* this age-old problem has plagued church organ pipes and even today poses problems with modern electronics as a shift to lead-free solders has left pure tin as lead's replacement for coating wire leads. Interestingly, this phenomenon does not occur when tin purity is below 99.9%.

In the manufacture of window glass, a vat of molten tin is used as a liquid bath upon which to float the glass. Known as the *Pilkington process,* after its inventor, this method of glassmaking is unmatched in producing flat panes of even thickness.

Table 2.47   Properties of Tin[*]

	Metric	English
Density	5.765 g/cm^3	0.2083 lb/in^3
Hardness, Brinell	3.9	3.9
Melting Point	231.968°C	449.543°F
Tensile Strength, Ultimate	220 MPa	31900 psi
Ultimate Yield Strength		

[*]Selected data from the MatWeb search database reprinted with permission from MatWeb, www.matweb.com.

2.7.6 Gold

Gold is a heavy, soft, precious metallic element, symbol Au, atomic number 79. It is the most malleable and ductile of all metals, and gold leaf is regularly produced with a thickness of only 0.000003 (three millionths) of an inch. Gold withstands oxidation and corrosion, holding its beautiful yellow color without the need for periodic polishing. Neither nitric nor hydrochloric acid will dissolve gold but, when combined, the two acids produce a solution called *aqua regia* that will readily do so.

Gold is found both in its native form and in ores. In its native form, gold is found in streams as flakes and small nuggets that have been eroded from their origin and washed downstream. As found in ores, gold is often combined with quartz and sulfur minerals, and a gold content of one gram per ton of ore is considered profitable.

History shows us that gold has been used since at least 2500 BC, when artisans used it for ornamental work and jewelry. The jewelers of today appreciate the many properties of gold no less than their ancient counterparts, and their work can be found adorning people of all walks. Other uses of gold include coinage, electrical contacts, dentistry work, and embroidery thread. Gold has been the standard for monetary systems until recent times, and although that has ended, gold is still easily and readily convertible to currency in most cultures. In fact, gold is often held to hedge against the inflation that plagues fiat currencies.

2.7.7 Platinum

Platinum is a silvery gray, dense, precious metallic element, symbol Pt and atomic number 78. It is malleable, ductile, and very resistant to corrosion, and it will not oxidize. It is found in its native form (known as polyxene) or in sperrylite, a sulfide ore.

Platinum is considered one of the most precious metals, and its price is often higher than that of gold. During World War II, it was considered a strategic metal, being needed for aircraft spark plugs, and all civilian use of the metal was made illegal and discontinued. Once the war ended, however, platinum once again was a favorite material of the jeweler due to its fine properties.

The main use for platinum today is inside automobile catalytic converters, where platinum takes carbon monoxide and unburned fuel vapors and reduces them into carbon dioxide and water vapor. It is still a favorite material for spark plugs, both in piston aircraft as well as in automobiles. And, of course, jewelers and their customers still love it for adornment.

2.7.8 Silver

Silver is a white, lustrous, metallic element that is exceptionally ductile and malleable. Known since antiquity as a precious metal, it is irreplaceable and unmatched even today due to its properties. Silver has the highest thermal and electrical conductivity of any metal and, being the whitest metal, is unparalleled as a reflector of light.

Its uses include jewelry, dental products, tableware, and coins. It is often applied as an electroplated finish on lesser metals and is indispensable in high-end mirrors and photographic processes. Silver has increased as a bullion metal and is available as ingots, bricks, and "rounds" to investors. The purity of bullion silver is usually 999/1000, while that for jewelry use is

925/1000, and known as *sterling silver*. The lack of purity is not intended to stretch out the silver but simply to reduce the softness of pure silver, which can bend or break easily in fine settings. The additive to sterling is copper, which hardens the metal sufficiently at its low percentage.

2.7.9 Titanium

Titanium is a strong, lightweight, corrosion-resistant metallic element (symbol Ti, atomic number 22) with a silvery-white color. It is not found in a native form but instead is extracted from two oxide ores: ilminite and rutile. Elemental, unalloyed titanium has the greatest weight-to-strength ratio of all metals, equaling mild steel in strength yet being 45 percent lighter. Its tensile strength, however, can easily be doubled or even tripled when alloyed with small amounts of aluminum, vanadium, chromium, and/or molybdenum. The most common titanium alloy, Ti_6Al_4V, contains 6 percent aluminum and 4 percent vanadium. Typically called grade 5, or simply "6-4," this metal has an ultimate tensile strength of 145,000 psi (1000 Mpa). Its strength and light weight, along with its chemical inertness and corrosion resistance, enable it to be crafted into artificial hips and other joints in the human body. Titanium is also used in aerospace and automotive components and sporting goods including bicycle frames and golf clubs. Being inert to most chemicals, it finds uses in the lab and field for caustic liquid storage and transport. Most of the titanium extracted from ore, however, goes into the manufacture of titanium dioxide and is used as a white pigment in products ranging from paint to toothpaste. Another by-product, titanium nitride (TiN), is a superhard coating often applied to wood and metal cutting tools to increase their performance.

Titanium is a difficult material to machine, especially if one is unaccustomed to it. For lathe turning, milling, and drilling, grade C2 carbide should be used when available. Ordinary high speed steel (HSS) tools can be used but must be kept sharp. Once HSS tools start to dull when cutting titanium, complete breakdown quickly results. Very slow cutting speeds are called for, but heavy cuts can be taken with good results when a flood coolant is used. The lower grades (1 through 4) offer the easiest machining, but at the sacrifice of inferior mechanical properties as compared to grade 5.

Part II

Plastics

3

Thermoplastics

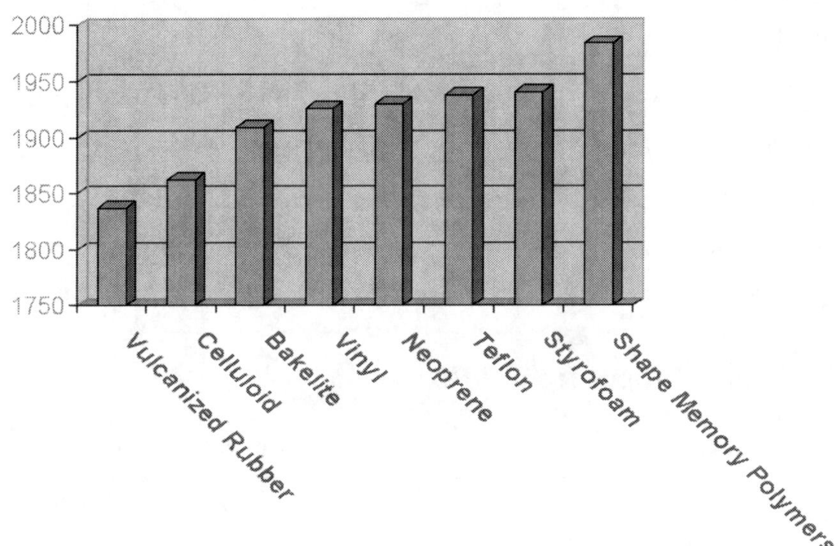

FIGURE 3.1 Invention timeline of various polymers.

3.1 ACRYLONITRILE BUTADIENE STYRENE (ABS)

ABS is a copolymer of acrylonitrile, butadiene, and styrene. ABS plastic is widely used because of its range of properties. It can be thermoformed into shape from sheet form, easily welded, and machined to close tolerances. ABS is highly impact resistant and possesses good strength and electrical insulating properties. It is a common plastic due to its availability and low cost. Different grades of ABS have been developed that possess special properties not normally found in standard grades; these include higher impact strength and heat resistance. Like most plastics, it is available as a resin for extrusion work as well as in common shapes for machining and thermoforming. ABS is available in a wide range of colors.

ABS plastics are used in business machines, telephone handsets, automobile instrument panels, and trim work. It is also used in piping to transport water, petroleum products, and some acids and alkalis. ABS is soluble in acetone and other ketones.

Table 3.1 Properties of ABS[*]

	English	Metric
Density	0.038 lb/in^3	1.04 g/cm^3
Hardness, Rockwell R	105	
Melting Point	221°F	105°C
Tensile Strength	6500 psi	44.8 MPa
Tensile Elongation at Break	25%	

[*]Selected data from the Material Selection Guide website, reprinted with permission from Boedeker Plastics (BPI), www.boedeker.com.

3.2 NYLON

Nylon was the first totally synthetic polymer, and it is most noted for its fibers, which are used in textiles and rope. By World War II, nylon had become a scarce commodity, as it was replacing silk in parachutes and was used to strengthen the tires of aircraft, heavy trucks, and automobiles. After the war, it became popular as a replacement for silk in women's stockings and for apparel that will not rot and grow mildew as natural fibers do.

Nylon is also used as an engineering plastic for gears, bearings, and general hardware. It is easily cast, extruded, and machined and has excellent mechanical properties. These include high impact and abrasion resistance as well as a very low coefficient of friction (0.15 for nylon on nylon).

Table 3.2   Properties of Nylon[*]

	English	Metric
Hardness, Rockwell R	92	
Melting Point	420°F	216°C
Tensile Strength	5800 psi	40 MPa

[*]Selected data from the Material Selection Guide website, reprinted with permission from Boedeker Plastics (BPI), www.boedeker.com.

3.3 POLYCARBONATE (LEXAN)

Polycarbonate is a tough, clear material used in many industrial, commercial, and domestic products. Its optical clarity makes it useful for a replacement of glass, with the added benefits of lower cost and higher impact resistance. In fact, polycarbonate is used as a bulletproof material for security-type glazing and aircraft windscreens. It has a higher refractive index than glass, which means that eyeglass lenses can be made thinner by using this material. Being plastic, polycarbonate also is lightweight, being about half as dense as even the lightest types of glass. Other useful attributes include good chemical resistance and excellent UV filtering properties. One major downfall to polycarbonate is its softness, which lends to poor scratch and abrasion resistance. To combat this, various hard coatings are either coextruded onto its surfaces during production or added afterward. Some of these coatings are nearly as hard as glass, which, while adding significant costs, still allows polycarbonate to be used to full benefit in harsh conditions. Polycarbonate is inexpensive, readily available, and easily worked by all common methods, making it a first choice for many products.

Polycarbonate's many uses include CDs and DVDs, which are produced in the millions worldwide. Other applications include machine guards, automobile lighting lenses, drinkware, and safety goggles. Polycarbonate is also produced in both transparent and opaque colors, allowing it to be used in products ranging from sunglass lenses to cellular phone cases.

Table 3.3   Properties of Polycarbonate[*]

	English	Metric
Density	1.2 lb/in^3	0.043 g/cm^3
Color	Transparent to colored	
Hardness, Rockwell M	70	
Tensile Strength (psi)	9500 psi	65.5 MPa
Tensile Elongation at Break	60%	

[*]Selected data from the Material Selection Guide website, reprinted with permission from Boedeker Plastics (BPI), www.boedeker.com.

3.4 POLYSTYRENE

Polystyrene is a rigid, transparent or colored plastic. It exhibits excellent flow properties, allowing extrusions and moldings with fine details. Polystyrene is an amorphous plastic, having no definite order of its long hydrocarbon chains. It is one of the more inexpensive and commonly used polymers. Two forms of polystyrene are regularly used: expanded foam and solid.

Expanded and extruded polystyrene foam is used for its insulative properties in many products including coffee cups, drink coolers, and structural panels. Other uses include packaging "peanuts" and life preservers. Styrofoam®, although a trademark of Dow Chemical, is a commonly used name for these types of foam.

Solid polystyrene is often extruded into sheets and formed into many common items including disposable plastic cups, break-apart auto and aircraft model sheets, CD cases, and electronic goods. Various formulations of solid polystyrene are produced having improved properties, including impact strength and toughness.

Table 3.4 Properties of Polystyrene[*]

	English	Metric
Density	0.043 lb/in^3	1.05 g/cm^3
Hardness, Rockwell M	75	
Tensile Strength	7500 psi	51.7 MPa
Tensile Elongation at Break	47%	

[*]Selected data from the Material Selection Guide website, reprinted with permission from Boedeker Plastics (BPI), www.boedeker.com.

3.5 POLYPROPYLENE

Polypropylene (PP) (Fig. 3.2) has two exceptional properties that set it apart from other consumer plastics: its resistance to heat and its ability to repeatedly flex without fatiguing. Food containers made from polypropylene can withstand the high heat in automatic dishwashers without warping, and most Tupperware® containers are in fact polypropylene. The anti-fatigue properties are useful for making hinges out of polypropylene and are known as "living" or "poly" hinges. Able to withstand literally millions of cycles without failure, such hinges are used in drinking cups and containers and are usually molded directly into the article.

Polypropylene is also used to make water-resistant fibers for indoor/outdoor carpeting. It also has many commercial and industrial uses including chemical resistant piping and storage vats, rope, and automobile trim.

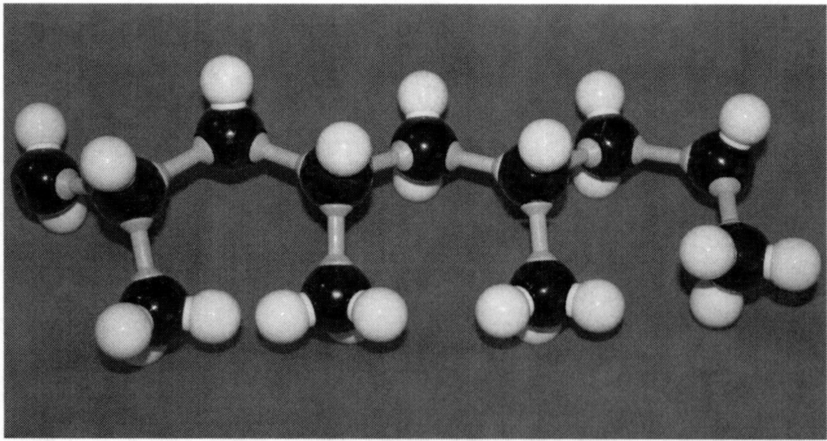

FIGURE 3.2 Structure of polypropylene, a simple hydrocarbon with alternating methyl (CH3) groups.

Table 3.5   Properties of Polypropylene[*]

	English	Metric
Color	Translucent	
Hardness, Rockwell R	95	
Melting Point	338°F	170°C
Tensile Strength	4500 psi	31 MPa

[*]Selected data from the Material Selection Guide website, reprinted with permission from Boedeker Plastics (BPI), www.boedeker.com.

3.6 POLYETHYLENE

Polyethylene (PE) is a long-chain thermoplastic polymer (Fig. 3.3). It is the most widely used plastic, being used for plastic bags, food and cosmetic packaging, and many engineered parts used in industry. Polyethylene molecules consist of long chains of carbon atoms with two hydrogen atoms attached to each. Different types of PE have more advanced molecules where separate chains branch off from the carbon atoms in the

FIGURE 3.3 Polyethylene, the simplest and most widely used polymer.

main chain. However, as advanced as these molecules are, compared to straight linear molecules, the branched molecules make a weaker plastic.

Polyethylenes are categorized by their molecular weight and density relative to each other. There are several varieties of PE, making it the most popular family of plastics.

Table 3.6   Properties of a Typical Polyethylene[*]

	English	Metric
Density	0.033 lb/in^3	0.92 g/cm^3
Hardness, Rockwell R	84	
Melting Point	160°F	71°C
Tensile Strength	1800–2200 psi	12.4–15.2 MPa

[*]Selected data from the Material Selection Guide website, reprinted with permission from Boedeker Plastics (BPI), www.boedeker.com.

3.7 LOW-DENSITY POLYETHYLENE

Low-density polyethylene (LDPE) is a soft, almost waxy plastic. It is a lightweight material, and it will float on water with a specific gravity of around 0.90. LDPE is not a very strong plastic but is very flexible and transparent, making it useful for plastic stretch wrap for both food and non-food items. It is often blow molded into squeeze bottles and pipettes for laboratory use, taking advantage of its chemical-resistant properties. At the thickness of these items, LDPE looses its transparency and becomes opaque. Other uses of LDPE include toys, plastic trim, and garbage bags.

Table 3.7   Properties of LDPE[*]

	English	Metric
Density	0.033 lb/in^3	0.92 g/cm^3
Melting Point	230°F	110°C
Tensile Strength	1800–2200 psi	12.2–15.2
Tensile Elongation at Break	600%	

[*]Selected data from the Material Selection Guide website, reprinted with permission from Boedeker Plastics (BPI), www.boedeker.com.

3.8 HIGH-DENSITY POLYETHYLENE

High-density polyethylene (HDPE) is one of the stronger polyethylenes due to its unbranched molecule chains. As compared to LDPE, it also has better high-temperature and chemical resistance and is somewhat harder. HDPE, being FDA and USDA approved for direct food contact, is often used in the food and beverage industry as blown containers and cutting boards. It is also one of the most recycled plastic materials, and countless otherwise discarded bottles and scraps make their way into plastic lumber, structural members, and decking. HDPE machines very well but is difficult to glue.

Table 3.8   Properties of HDPE[*]

	English	Metric
Density	0.035 lb/in^3	0.95 g/cm^3
Melting Point	230°F	110°C
Tensile Strength	4600 psi	31.7 MPa

Table 3.8   Properties of HDPE[*] *(cont.)*

	English	Metric
Tensile Elongation at Break	900%	

[*]Selected data from the Material Selection Guide website, reprinted with permission from Boedeker Plastics (BPI), www.boedeker.com.

3.9 ULTRA-HIGH-MOLECULAR-WEIGHT POLYETHYLENE

Ultra-high-molecular-weight polyethylene (UHMWPE) is a very durable, wear-resistant, and low-friction plastic. It has the highest impact resistance of any plastic and, like all of the polyethylenes, has good chemical resistance. UHMWPE also performs excellent in cold temperatures where other plastics would become brittle and crack. As such, it is used in snowmobile tracks and running surfaces. Other uses include gears, sprockets, and bearings that can replace metals in many applications. The low-friction and wear-resistance properties allow UHMWPE to be used as a substitute for ice in skating rinks, thus allowing a rink to be installed by the sheet in any area.

UHMWPE fibers are another common form of this plastic. The fibers possess the same toughness and anti-friction properties of the solid polymer and are used in body armor, fishing line, parachute cordage, and climbing equipment.

Table 3.9   Properties of UHMWPE[*]

	English	Metric
Density	0.034 lb/in^3	0.93 g/cm^3

Table 3.9   Properties of UHMWPE[*] *(cont.)*

	English	Metric
Melting Point	280°F	138°C
Tensile Strength	4000 psi	27.6 MPa
Tensile Elongation at Break	140%	

[*]Selected data from the Material Selection Guide website, reprinted with permission from Boedeker Plastics (BPI), www.boedeker.com.

3.10 POLYMETHYLMETHACRYLATE (ACRYLIC)

Polymethylmethacrylate (PMMA) thermoplastic is transparent and often used in optical applications. It is a hard, rigid plastic that is half the weight, has better light transmission, and is more shatter resistant relative to glass. When compared to polycarbonate, however, PMMA's impact resistance is extremely inferior. PMMA also machines poorly due to its brittleness and is easily chipped and broken. When cost is an issue, however, PMMA can be chosen due to its lower price, provided that its weaknesses are taken into consideration. PMMA also has slightly better light transmission than polycarbonate, which makes itself evident in thicker sections.

Besides its uses in glazing, PMMA is used in acrylic paint, where the powdered acrylic polymer is mixed with a water base. PMMA is also used in fiber optic cables due to its transparency and refractive index.

Table 3.10   Properties of PMMA[*]

	English	Metric
Density	0.043 lb/in^3	1.18 g/cm^3
Hardness, Rockwell M	80–100	
Melting Point	265–285°F	130–140°C
Tensile Strength	8,000–11,000 psi	55.2–75.8 MPa
Tensile Elongation at Break	2%	
Refractive Index	1.48–1.50	

[*]Selected data from the Material Selection Guide website, reprinted with permission from Boedeker Plastics (BPI), www.boedeker.com.

3.11   POLYTETRAFLUOROETHYLENE (TEFLON®)

Polytetrafluoroethylene (PTFE) plastic is known mainly for its low-friction qualities. It is the slipperiest plastic known, having a friction coefficient of 0.04 when testing PTFE on PTFE. As such, it is used for bearing surfaces on machinery, sliding surfaces on conveyors, and of course the nonstick surface on cookware. Its slippery surface makes it resistant to contaminant buildup and, interestingly, it is the only material that a gecko lizard cannot climb. PTFE has other useful properties, though, including great chemical resistance and insulative qualities. Gore-Tex®, a trademarked name for a PTFE fiber, is used in wet weather wear to allow perspiration to escape from within while repelling water from the outside.

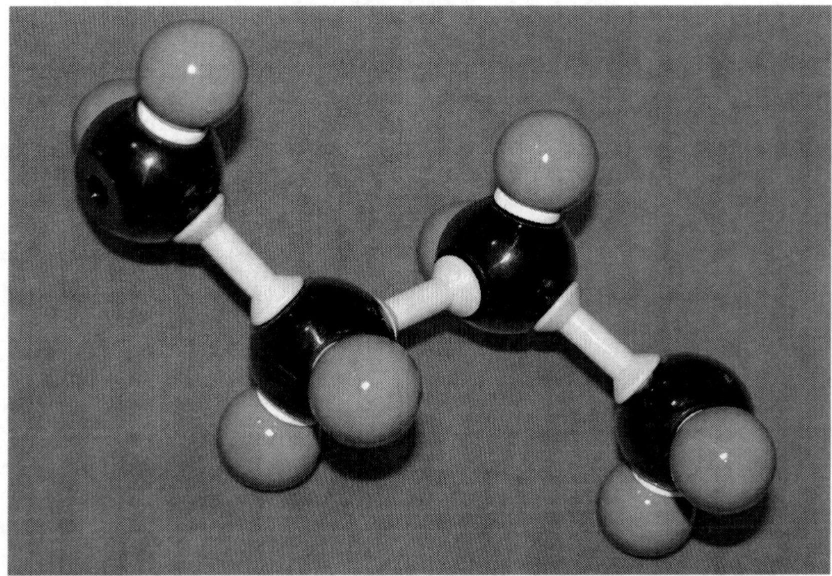

FIGURE 3.4 Polytetrafluoroethylene, also known as PTFE or Teflon, is similar in structure to polyethylene but has fluorine atoms in place of hydrogen alongside of carbon.

Table 3.11   Properties of PTFE[*]

	English	Metric
Density	0.078 lb/in^3	2.16 g/cm^3
Melting Point	635°F	335°C
Tensile Strength	3900 psi	26.9 MPa
Tensile Elongation at Break	300%	

[*]Selected data from the Material Selection Guide website, reprinted with permission from Boedeker Plastics (BPI), www.boedeker.com.

3.12 POLYVINYLIDENE CHLORIDE

Polyvinylidene chloride (PVDC) resins are used mainly in packaging films for both food and non-food items, having superior properties over hydrocarbon films. PVDC films are extremely transparent and glossy and provide excellent barriers against oxygen, water, odors, and chemicals. They also are able to provide strong hermetic seals that protect their contents. Besides being used alone, PVDC films are also used in conjunction with other plastics as an added barrier in items such as blister packs for pills and medical waste bags. Saran™ Wrap, created by Dow Chemical, was a common PVDC consumer film used in the kitchen. Due to concerns about its chlorine makeup, Dow changed the formula, and currently Saran Wrap is made from more inert hydrocarbons, at the cost of the original material's effectiveness.

Table 3.12　Properties of PVDC[*]

	English	Metric
Density	0.064 lb/in^3	1.77 g/cm^3
Melting Point	332°F	166°C
Tensile Strength	6300 psi	43.4 MPa
Tensile Elongation at Break	50%	

[*]Selected data from the Material Selection Guide website, reprinted with permission from Boedeker Plastics (BPI), www.boedeker.com.

3.13　POLYETHYLENE TEREPHTHALATE

Polyethylene terephthalate (PET, PETE, polyester, Dacron[®], Mylar[®]) is a strong, lightweight, and inexpensive plastic used mainly as water and soft drink bottles (Fig. 3.5). When used in drinkware, PET is transparent and chemically inert. This form

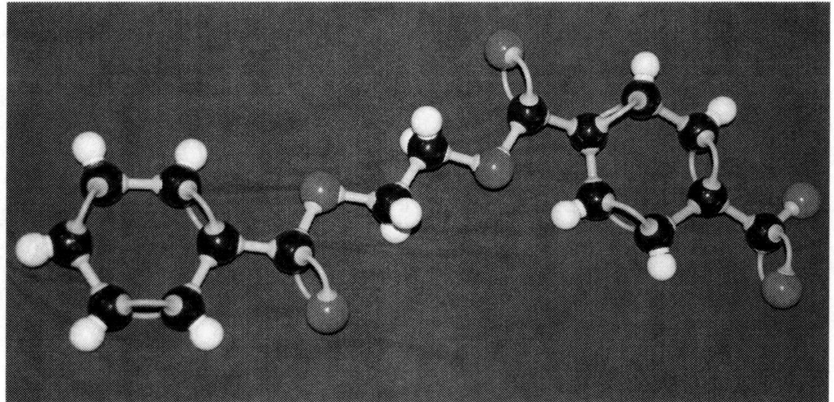

FIGURE 3.5 Structure of polyethylene terephthalate. Also known as PET, polyester, and Mylar. Note the recurring benzene rings.

of PET is also blow-molded or vacuum formed into packaging for many consumer products. PET resin also makes up the familiar polyester fibers found in garments, furniture, and blankets. PET films (Mylar®) are used in graphics displays and packaging, and as electronic media tape when magnetically coated. This form of PET is specially drawn during manufacturing to obtain high levels of strength and stiffness. PET is often recycled into carpet and textile fibers as well as bottles and jars.

Table 3.13   Properties of PET[*]

	English	Metric
Density	0.051 lb/in^3	1.41 g/cm^3
Hardness, Rockwell M	93	
Melting Point	491°F	255°C
Tensile Strength	12,400 psi	85.5 MPa
Tensile Elongation at Break	20%	

[*]Selected data from the Material Selection Guide website, reprinted with permission from Boedeker Plastics (BPI), www.boedeker.com.

3.14 POLYVINYL CHLORIDE

Polyvinyl chloride (PVC) plastic is most commonly seen as potable water and sewage piping in homes, businesses, and industrial plants (Fig. 3.6). PVC pipe is chemically stable, flame resistant, easy to work with, and inexpensive. It also is resistant to many chemicals and oils, furthering its use for piping and storage vats for these materials. PVC is the vinyl siding used on house exteriors and is further used in products ranging from automobile trim to raincoats and shower curtains. For these softer materials, plasticizers are added to give suppleness to an otherwise harder plastic.

Table 3.14 Properties of PVC[*]

	English	Metric
Color	Amber colored	
Hardness, Rockwell R	115	
Tensile Strength	7500 psi	51.7 MPa

[*]Selected data from the "Plastics Database," reprinted with permission from Plasticsusa.com, www.plasticsusa.com.

FIGURE 3.6 Structure of polyvinyl chloride (PVC). Similar to polyethylene, though having an alternating chlorine atom replacing hydrogen.

3.15 POLYETHER ETHER KETONE

Polyether ether ketone (PEEK) is a high-performance engineering plastic noted for its high-temperature properties and chemical resistance. It can safely withstand constant temperatures of 480°F without failure or softening, and it has a melting temperature of 644°F. PEEK is a tough, strong plastic with excellent wear and abrasion resistance and is often used for bearing surfaces. It is available in four main types: unfilled, glass-filled, carbon-filled, and bearing grade.

Glass-filled PEEK has better deformation resistance at high temperatures than unfilled PEEK and also has better tensile and compression strengths. Carbon-filled PEEK has higher tensile and compression strengths than glass-filled, has a higher operating temperature (500°F), is stiffer, and has a lower coefficient of friction. Bearing-grade PEEK is reinforced with carbon fiber, graphite, and PTFE (Teflon[R]). It has the lowest friction coefficient and highest wear and abrasion resistance, and it is the easiest machining PEEK plastic.

Table 3.15   Properties of PEEK[*]

	English	Metric
Density	0.047	1.31
Hardness, Rockwell M	100	
Melting Point	644°F	340°C
Tensile Strength	16,000 psi	110.3 MPa
Tensile Elongation at Break	20%	

[*]Selected data from the Material Selection Guide website, reprinted with permission from Boedeker Plastics (BPI), www.boedeker.com.

3.16 THERMOPLASTIC POLYURETHANE

As the one class of polyurethanes that are not thermosets, thermoplastic polyurethanes (TPUs) have the advantage over thermoset polyurethanes in their ability to be formed as a thermoplastic; i.e., through a conventional extrusion process. Like their thermoset cousins, TPUs are available in a wide range of grades from soft elastomers to rigid bearing plastics, and after processing they have similar properties to regular thermosetting polyurethanes. Glass and/or carbon fibers are added to specific grades to increase desired properties. The ease in fabricating parts from TPUs as compared to thermoset polyurethanes has made them increasingly popular for many products including bearings, wear plates, mechanical shock absorbers, and all types of flexible and rigid foams. Being thermoplastic, TPUs are also recyclable.

Table 3.16   Properties of TPUs[*]

	English	Metric
Density	0.051 lb/in^3	1.42 g/cm^3
Tensile Strength	10,000 psi	68.9 MPa
Tensile Elongation at Break	4%	

[*]Selected data from the Material Selection Guide website, reprinted with permission from Boedeker Plastics (BPI), www.boedeker.com.

3.17 LIQUID CRYSTAL POLYMER

Liquid crystal polymers (LCPs, e.g., Xydar® and Vectra®) are named for their maintained crystalline structure in a liquid state. These crystals, made up of long spaghetti-like molecules, are self-aligning in the direction of flow during processing. LCPs have very high heat resistance due to the material's high melting temperature. This high heat melt, in addition to the molecular alignment of the resin, delivers a very low-viscosity product that, when delivered through an injection molder, can be forced into extremely thin wall sections and very intricate shapes. The crystalline alignment gives these thin walls high strength and toughness that would be impossible to achieve with other resins. Other properties of LCPs are chemical resistance and very low thermal expansion. Although LCP resins are processed at very high temperatures relative to other polymers, similar sized LCP parts do in fact solidify in about half the time, enabling faster parts cycling. Uses include electronic components that will endure solder heat and specially shaped products requiring strong, thin walls. LCP resins are available alone or with added fiberglass and/or carbon fibers for added strength.

Kevlar® and similar fabrics and fibers are made from LCPs. The longitudinal alignment of the molecules gives this material its great strength, and it is used for many purposes, including body armor and fishing line.

4

Thermosets

Thermoset plastics undergo a non-reversible hardening when they cure, either through a thermal or chemical reaction. Unlike thermoplastics, they cannot be remelted upon heating, and, when brought to a high enough temperature, will decompose and eventually burn. The polymer chains cross-link to one another with strong intermolecular bonds forming massive 3-D molecules. Once cured, thermosets are generally harder and stiffer than most thermoplastic materials. To increase the material's strength, glass fibers or metallic powders are often added to the mixture before curing. Added graphite will increase the material's lubricity for use as bearing surfaces, while various pigments and plasticizers are also used.

4.1 EPOXY

Epoxy, or polyepoxide, is a thermoset resin that cures and hardens when mixed with a proper catalyst. Epoxies are used as adhesives, electrical insulators, and composite structural shapes. Epoxy adhesives are used both in industry and in households to secure materials ranging from aircraft structures to a broken coffee cup handle. It has excellent bonding properties, and individual formulas are made for specific materials, including various metals, woods, and plastics. Being electrically insulative, epoxy is used to produce circuit boards and molded components in electrical motors, transformers, and similar items. Epoxy resin is often molded with fiberglass sheeting to produce larger items, including boat hulls, automotive bodies, and storage tanks. Besides fiberglass, carbon fibers and Kevlar® are also used as strengtheners.

Epoxy formulas vary according to thickness and cure time depending on desired properties and working time needed. They are also formulated with conductive metal powders that can be used as a cold cure solder for areas that are heat sensitive. Epoxy is also used as a tough coat paint for high-wear items such as floors and as a corrosion-proof finish on steel. These finishes can also tolerate higher temperatures than conventional paints and, once applied, are very durable—often for the life of the product.

4.2 MELAMINE RESIN

Melamine resin is a thermoset produced from the combination of the compound melamine and formaldehyde. Melamine resin is hard, durable, and very commonly used for inexpensive kitchen utensils and tableware. The familiar name Formica® is a lamination of melamine resin combined with paper or fabric binders and used for countertops. Melamine cups and dishes are popular for camping and other temporary or occasional use. They are not microwave safe, however, and tend to stain over time. Older melamine utensils and dinnerware items have recently become favorites of collectors, who pay a high premium for rare or colorful pieces. Other uses of melamine resin include dry erase boards and plastic furniture.

4.3 PHENOLIC THERMOSETS

Phenolic thermosets, or simply phenolics, are a group of phenol/formaldehyde resins commonly combined with paper, glass, or cotton fibers and cured under pressure to the desired form. The form may be a finished product or an extruded shape to be machined later. Phenolic plastics are hard, rigid, and tough to the point of being used in the manufacture of billiard balls. They are also resistant to corrosion and fire and are used in marine environments aboard ships and oil rigs for pipe sections, firewalls, and general construction elements. Another major use of these resins is in the manufacture of electronic printed circuit (PC) boards, which are made with either glass or cotton fibers. Phenolic PC boards are excellent electrical insulators, are heat resistant, and can be made inexpensively. In more intricate boards, complex circuitry is laminated between multiple layers of phenolic in crisscross patterns that would not be possible with single-layer boards. Bakelite plastics are made of phenol/formaldehyde resins that use wood as a binder. They have been in use for over 100 years and vary from modern phenolics only in terms of the binding materials used in their makeup.

4.4 BAKELITE/CATALIN

Bakelite is a thermosetting plastic made from a carbolic acid and formaldehyde resin with a pulverized wood filler. It was one of the first synthetic plastics, having been produced since the early 1900s. When vigorously rubbed, scraped, or burned, it gives off a telltale formaldehyde odor. In its early days, Bakelite was cast into many products that were formerly made from wood, glass, and metal, including radio chassis, children's toys, and electrical insulators. In 1927, the patent for Bakelite expired, and the Catalin Company snapped up the formula. Catalin changed things by leaving out the wood filler and adding vibrant colors to this new formula, customarily swirling multiple colors together into single batches. Each lot was cast into rods, blocks, and sheets and then cut and carved into individual bangles, trinkets, and beads. Catalin jewelry was an instant success and remains so even today due to its style and appeal. There are many plastic replicas of Catalin, but none has the density and rich properties of the original.

4.5 ELASTOMERS

4.5.1 Butyl Rubber

Butyl is a synthetic rubber made from a copolymer of isobutylene and isoprene. The two main features of butyl are its impermeability to liquids and gasses and its excellent flexibility, making it indispensable for tire inner tubes, sports ball bladders, and chemical gas masks. It is also used as a general sealant for

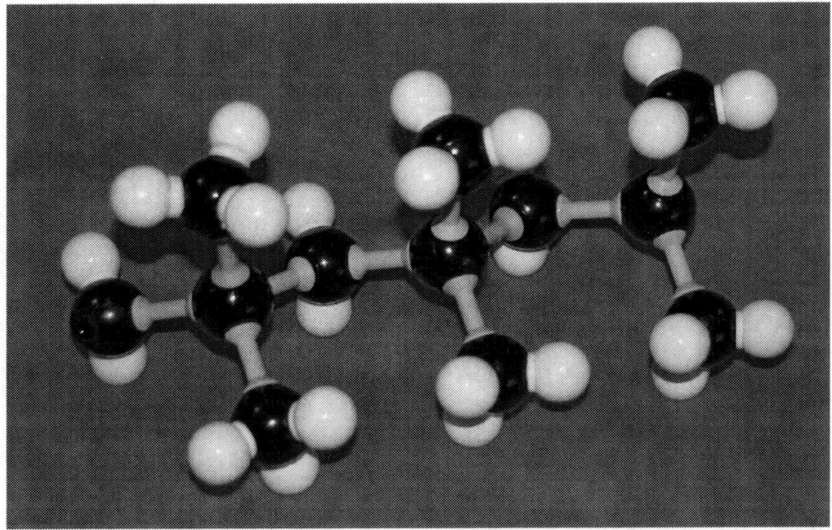

FIGURE 4.1 Butyl rubber structure—similar to polypropylene but having double methyl (CH3) groups alternating between CH2 groups.

use on metals, wood, mortar, and glass. Sold in cartridges, cans, and drums, butyl rubber caulk is an air-drying solvent base form of butyl and is used to seal joints in truck trailers, shipping containers, and railroad cars. It is also used around the home for gutters, downspouts, and flashings. In addition, there are butyl tapes available that are used as gaskets for glass-to-metal seals, including automotive and aerospace windshields. A novel use for butyl is chewing gum base. As compared to natural chicle, butyl gum is less expensive, stays soft on the shelf for a longer time, and retains its flavor longer.

4.5.2 Natural Rubber

Natural rubber is a product made from the latex sap of several trees, most importantly *Hevea brasiliensi*. The rubber tree is tapped by cutting diagonal slits across the trunk that not only open up the tree but also direct the sap into collection pails. After the sap is collected, it is mixed with a coagulator and rolled out into sheets, after which it is ready for export. This type of raw rubber is known as *latex rubber* and is used for items ranging from rubber gloves to balloons and condoms. Most rubber undergoes the process of vulcanization, which makes the rubber harder, stronger, and element resistant. In vulcanization, sulfur is added to latex, and the resultant material is compressed at a high temperature. At this high pressure/temperature, the sulfur acts as a bridge between the independent latex molecules, and cross-linking (curing) occurs.

Natural rubber is chemically called *polyisoprene* and, when produced synthetically, it is known as *synthetic natural rubber.* Polyisoprene has all of the desirable properties of natural rubber and is not dependent on successful latex farming and transportation.

The largest market for natural rubber is in the automotive and aircraft tire industry. Other uses of rubber include athletic shoes, adhesives, and various tubing. Many items formerly made of natural rubber are now made from different synthetics having superior properties.

4.5.3 Nitrile Rubber (Buna-N)

Nitrile rubber is a thermoset copolymer of acrylonitrile and butadiene capable of resisting fuels, oils, and many other chemicals at temperatures ranging from $-40°F$ to over $120°F$

(–40 to 49°C). By altering the ratio of the two polymers, nitrile rubber's properties can be balanced by elasticity and chemical resistance to fit the requirement. One of the most common uses of nitrile rubber is in the automotive industry for the manufacture of fuel, vacuum, and coolant hoses and seals. In general industry, nitrile rubber O-rings, hydraulic hoses, and pump seals are common. When strengthened with fiberglass strands, nitrile is used for vee and flat belts for all types of machinery. Nitrile rubber gloves are used in the medical field as a replacement for latex gloves to alleviate the common latex allergy among its users.

4.5.4 Polychloroprene (Neoprene)

Polychloroprene, or neoprene, was initially made by Dupont in 1930 and was the first mass-produced synthetic rubber. It has enhanced properties over natural rubber, including abrasion resistance, chemical resistance, and the ability to remain stable in temperatures ranging from –40 to 125°F (–40 to 52°C). Polychloroprene can be hot bonded to steel through a steam vulcanization process and is used to coat pipelines for both water and underground use. It is also bonded with fiberglass belts for use in pliable tubing and belting. A popular consumer use of neoprene is in wetsuits. The neoprene resin is nitrogen foamed, rolled into sheets, and sewed into shape. The micro nitrogen bubbles provide both buoyancy and thermal insulation for the diver or surfer. In the non-foamed variety, neoprene is used for gloves, snow boots, and fisherman's waders.

4.5.5 Silicone Rubber

Silicone rubber (Fig. 4.2) is a synthetic elastomer noted for its resistance to extreme temperatures and its weathering abilities. It is made from a silicon-based polymer strengthened with fillers, and it is available in either factory-cured shapes or *room-temperature vulcanization (RTV)* gels.

RTV gels are used for many projects in the home and for industry, and they are impervious to water and many chemicals, adhere well to most surfaces, and have good tear strength. They are used as sealants in fish tanks, bathtubs, and sinks as well as in exterior applications. RTV gels also make excellent adhesives and gaskets for automotive and other mechanical works. RTV silicones are available in two types: RTV1 and RTV2. RTV1 is the consumer-familiar type that comes in a tube, ready

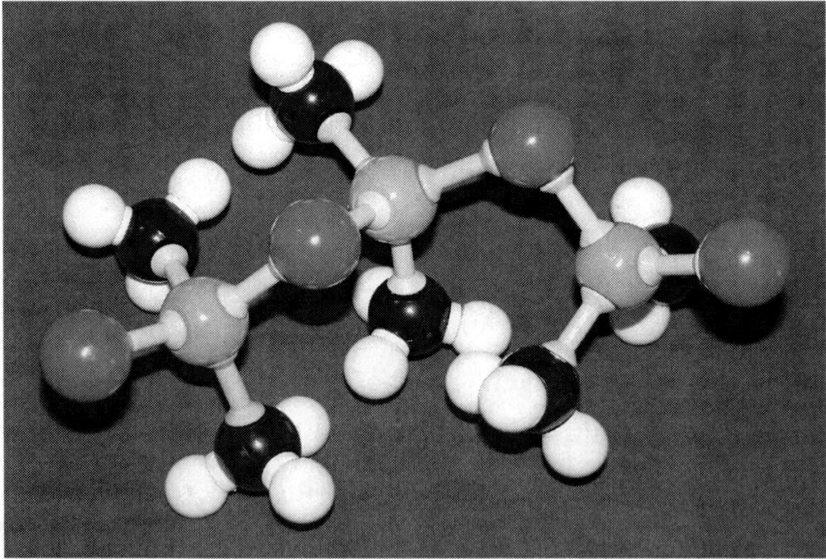

FIGURE 4.2 The structure of silicone rubber shows a silicon/oxygen trunk with branching methyl (CH3) groups.

to use. RTV2 silicone is a two-part system that must be mixed before use. This type is used for heavier cross sections and can cure from the inside out upon mixing. Its main use is in mold making. Silicone molds are tough, easy to make, and capable of withstanding most castable materials.

Factory vulcanized shapes are varied and many. Due to the high temperature tolerance of silicone, it is used as flexible bakeware in the home kitchen. Automotive, marine, and aircraft trim strips and windshield wiper blades are also molded and extruded from silicone.

Part III

Wood

5
Hardwoods

5.1 AMERICAN ASH (*FRAXINUS AMERICANA*)

The American Ash is found in eastern North America from Nova Scotia to Florida. It is a large, deciduous tree growing to 80 ft in height and having a trunk diameter of 2 ft. In addition to its importance as wood, ash also is used as an ornamental, producing burgundy to scarlet autumn foliage. Its bark is gray with a uniform diamond furrowing. Ash trees are susceptible to the emerald ash borer beetle.

As a wood, ash has the greatest strength-to-weight ratio, making it useful as the number one material for baseball bats. Its sapwood is light tan, darkening to brown in the heartwood. It is straight grained, very hard, and shock resistant, making it useful as flooring. Other uses include tool handles and furniture. Ash is one of the best bending woods when steamed, having very long grain structures. It is easily worked with hand and power tools and glues well.

Table 5.1   American Ash, Green Wood Properties[*]

	Metric	English
Weight	769 kg/m^3	48 lb/ft^3
Modulus of Rupture	66.192 MPa	9600 psi
Janka Hardness	4270.08 N	960 lbf
Sheer	9.308 MPa	1350 psi

[*]Source: U.S. Department of Agriculture, Forest Service, General technical report FPL-GTR-83, *Hardwoods of North America* (1995), by Harry A. Alden.

5.2 AMERICAN BASSWOOD *(TILIA AMERICANA)*

The American Basswood is found in the American Northeast and westward to Minnesota. It can grow to 120 ft and has a long, straight trunk that can reach 5 ft in diameter. Basswood leaves are symmetrical and heart shaped with serrated edges. Its bark varies from brown to gray with deep vertical plates. Cordage can be made from the bark of its branches due to its long interwoven fibers. When a basswood is cut down, several shoots grow from the trunk and propagate into individual trees.

The sapwood of the basswood is cream colored, and its heartwood is reddish brown with darker streaks. The wood is soft and light, with even texture, and is a favorite of woodcarvers. It lacks notable grain and can be cut against and with the grain equally well. Other uses of basswood include low-grade mass-produced furniture, plywood fill, and pulp stock.

Table 5.2   American Basswood, Green Wood Properties[*]

	Metric	English
Weight	673 kg/m^3	42 lb/ft^3
Modulus of Rupture	5000 psi	34.475 MPa
Janka Hardness	56.2 N	250 lbf
Sheer	600 psi	4.137 MPa

[*]Source: U.S. Department of Agriculture, Forest Service, *Tropical Timbers of the World* (1980), by Martin Chudnoff.

5.3 AMERICAN BEECH (*FAGUS GRANDIFOLIA*)

Beech trees are found in North America, Europe, and Asia. The American Beech can be found in southeastern Canada and the eastern United States. It grows to 120 ft in height, with a truck diameter of 4 ft. It has a smooth, gray bark. Animals eat beechnuts as part of the forest mast, and people also enjoy them for their woodsy flavor.

Beech sapwood is white with a reddish tinge, while its heartwood is reddish brown. Beechwood is hard and shock resistant, making it useful in flooring. Its straight grain, strength, and even texture make it suitable for dimensional lumber, veneer, pallets, and furniture. Although it is difficult to work with hand tools, it cuts well with machinery.

Table 5.3 American Beech, Green Wood Properties[*]

	Metric	English
Weight	865 kg/m^3	54 lb/ft^3

Table 5.3 American Beech, Green Wood Properties* *(cont.)*

	Metric	English
Modulus of Rupture	59.297 MPa	8600 psi
Janka Hardness	3780.80 N	850 lbf
Shear	8.894 MPa	1290 psi

*Source: U.S. Department of Agriculture, Forest Service, *Tropical Timbers of the World* (1980), by Martin Chudnoff.

5.4 AMERICAN CHESTNUT (*CASTANEA DENTATE*)

American Chestnut is an important deciduous tree of the North American Appalachian region. It can grow to 120 ft tall with a trunk diameter of 7 ft. In the first half of the twentieth century, the American Chestnut was almost wiped out by a blight that was brought over with imported Asian Chestnut trees. This same blight continues to keep the chestnut populations down in the U.S.A., but resistant varieties are being developed. The nut is both a favorite holiday food and a once important forest feed.

Chestnut wood has white sapwood and gray/brown heartwood that darkens with age. The wood is coarse, strong, and straight grained with medium hardness. It grows faster than oak and once was important for barn boards, utility poles, and split-rail fences. Other uses included railroad ties, firewood, and mine timbers. Today, due to the low availability of new chestnut wood, older boards and beams are being reused for many items. Chestnut is easily worked with hand tools and machinery and is being used for furniture and cabinetry. Chestnut glues easily but is difficult to nail without splitting.

Table 5.4 American Chestnut, Green Wood Properties[*]

	Metric	English
Weight	881 kg/m^3	55 lb/ft^3
Modulus of Rupture	38.612 MPa	5600 psi
Janka Hardness	1868.16 N	420 lbf
Shear	5.516 MPa	800 psi

[*]Source: U.S. Department of Agriculture, Forest Service, General technical report FPL-GTR-83, *Hardwoods of North America* (1995), by Harry A. Alden.

5.5 AMERICAN ELM *(ULMUS AMERICANA)*

The American Elm is found in eastern North America, from Newfoundland to Florida. It can reach a height of over 100 ft, with a trunk diameter of over 3 ft. Its leaves are serrated, oblong, and teardrop shaped with deep veining and dark green color. The bark of the elm is dark gray with deep ridges and furrows. The shape of the elm tree is unique, having the appearance of an inverted cone. Many elm trees have succumbed to the fungal Dutch elm disease, originally brought to North America from the Netherlands in 1828 by the elm bark beetle.

Elm sapwood is nearly white, while its heartwood is reddish to light brown. Due it its interlocking grain, it can be successfully steam bent without splitting and cracking. It has a pleasant grain structure and is used in furniture and cabinetry. Elm is also used as veneer over lesser woods and for specialty boxes.

Table 5.5   American Elm, Green Wood Properties[*]

	Metric	English
Weight	865 kg/m^3	54 lb/ft^3
Modulus of Rupture	49.644 MPa	7200 psi
Janka Hardness	2757.76 N	620 lbf
Shear	8.963 MPa	1300 psi

[*]Source: U.S. Department of Agriculture, Forest Service, General technical report FPL-GTR-83, *Hardwoods of North America* (1995), by Harry A. Alden.

5.6 AMERICAN HORNBEAM *(CARPINUS CAROLINIANA)*

The American Hornbeam ranges mainly in the eastern United States, although it is also found in central Mexico. It is a small, spindly tree with crooked growth and green/gray bark that rarely grows beyond 20 ft tall. It often grows in the shadows of larger deciduous trees in bottomlands and along rivers and streams.

Also known as ironwood, hornbeam is tough, hard, and heavy. Its sapwood is white, while its heartwood varies from white to pale yellow and tan. As it normally does not split and crack, hornbeam is useful for turning bowls and dishes when it can be found large enough. Its hardness, although making it difficult to work, makes it useful for tool handles, levers, and mallets.

Table 5.6   American Hornbeam, Green Wood Properties[*]

	Metric	English
Weight	849 kg/m^3	53 lb/ft^3
Modulus of Rupture	46.866 MPa	6800 psi
Janka Hardness	4181.12 N	940 lbf
Shear	7.998 MPa	1160 psi

[*]Source: U.S. Department of Agriculture, Forest Service, General technical report FPL-GTR-83, *Hardwoods of North America* (1995), by Harry A. Alden.

5.7 AMERICAN WHITE BIRCH/PAPER BIRCH (*BETULA PAPYRIFERA*)

The American Birch is a common deciduous tree found throughout Canada and the northern United States. It grows up to 70 ft tall with a 2-ft diameter trunk and has bright, papery white bark that can be peeled off in large sheets. It was at times used as writing paper. Its leaves are ovate, serrated, and from 2 to 4 in long. They are shade intolerant, preferring open areas. Birch will often reseed a fire burned area very quickly, forming large stands.

Birchwood is commonly used as firewood, delivering a hot long-lasting fire. It splits easy, which is good for the firewood cutter, but this property also makes nailing difficult without pre-drilling. Birch has straight grain and a fine, uniform texture. Its sapwood is white to near white while its heartwood is darker brown. It is easy to work with hand tools, making it useful for the hobbyist. Birch lumber is used as a veneer, in plywood, and in furniture. It also is used in cooperage and as pulpwood, and its sap is used in medicines, syrups, and soft drinks.

Table 5.7   American Birch, Green Wood Properties[*]

	Metric	English
Weight	801 kg/m^3	50 lb/ft^3
Modulus of Rupture	44.128 MPa	6400 psi
Janka Hardness	2,490.88 N	560 lbf
Shear	6.343 MPa	0.84 psi

[*]Source: U.S. Department of Agriculture, Forest Service, General technical report FPL-GTR-83, *Hardwoods of North America* (1995), by Harry A. Alden.

5.8 ASPEN

5.8.1 Bigtooth Aspen *(Populus grandidentata)* and Quaking Aspen *(Populus tremuloides)*

The aspens are found from Alaska through Canada, and south to Mexico. They prefer well lit, high altitudes. Although aspen occasionally lay seed, they prefer to propagate through their roots, which gives them the ability to survive forest fires. Aspen can grow to over 120 ft in height with 4-ft trunks. The bark varies considerably in both color and texture from white to brown and smooth to deeply furrowed. The leaves of the aspen turn golden in autumn and, when windblown, produce the pleasing rustling sound that they are famous for.

Aspen sapwood is white and blends into a light brown heartwood. The wood is very light, straight grained, and soft. Besides being a popular pulpwood, aspen is used in matchsticks, tongue depressors, crates, and boxes. Its softness, low shrinkage, and straight grain make it an easily carved wood.

Table 5.8 Aspen, Green Wood Properties[*]

	Metric	English
Big Tooth Aspen		
Weight	689 kg/m^3	43 lb/ft^3
Modulus of Rupture	37.233 MPa	5.40 psi
Janka Hardness	1645.76 N	370 lbf
Shear	5.033 MPa	0.73 psi
Quaking Aspen		
Weight	689 kg/m^3	43 lb/ft^3
Modulus of Rupture	35.165 MPa	5.10 psi
Janka Hardness	1334.40 N	300 lbf
Shear	4.551 MPa	0.66 psi

[*]Source: U.S. Department of Agriculture, Forest Service, General technical report FPL-GTR-83, *Hardwoods of North America* (1995), by Harry A. Alden.

5.9 BLACK CHERRY (*PRUNUS SEROTINA*)

Black Cherry is found in the eastern United States from Maine to Florida and also occurs commonly in Mexico. It can grow to over 100 ft with a trunk diameter of 5 ft, but most trees grow shrubby and are not suitable for use as lumber. On mature trees, the bark is dark gray to brown with deep breaks. On younger specimens, it is nearly smooth, with thinner and shallower furrows. Cherry fruit is popular with both people and birds, who often compete for it.

Cherry sapwood is light yellow, while its heartwood is brown. Upon exposure, the heartwood darkens to the deep reddish brown that is so prized by furniture makers. Cherry wood is lightly aromatic and is strong, uniform textured, and generally straight grained. It is easily worked, glues well, is dimensionally stable, and finishes nicely. Besides being used in fine furniture, cherry is used for upscale toys, piano actions, and musical instruments. It is also a favorite wood for veneer and paneling due to its warm grain and deep colors.

Table 5.9   Black Cherry, Green Wood Properties[*]

	Metric	English
Weight	45 lb/ft^3	721 kg/m^3
Modulus of Rupture	55.160 MPa	8.00 psi
Janka Hardness	2935.68 N	660 lbf
Shear	7.791 MPa	1.13 psi

[*]Source: U.S. Department of Agriculture, Forest Service, General technical report FPL-GTR-83, *Hardwoods of North America* (1995), by Harry A. Alden.

5.10 BLACK WALNUT *(JUGLANS NIGRA)*

The Black Walnut (Fig. 5.1) is found in the eastern United States but more commonly in the American Midwest. The trees can reach a height of over 100 ft with a diameter of over 3 ft. Its bark is dark and deeply furrowed, and its compound leaves can reach over 2 ft in length. Its twigs have a chambered pith that is

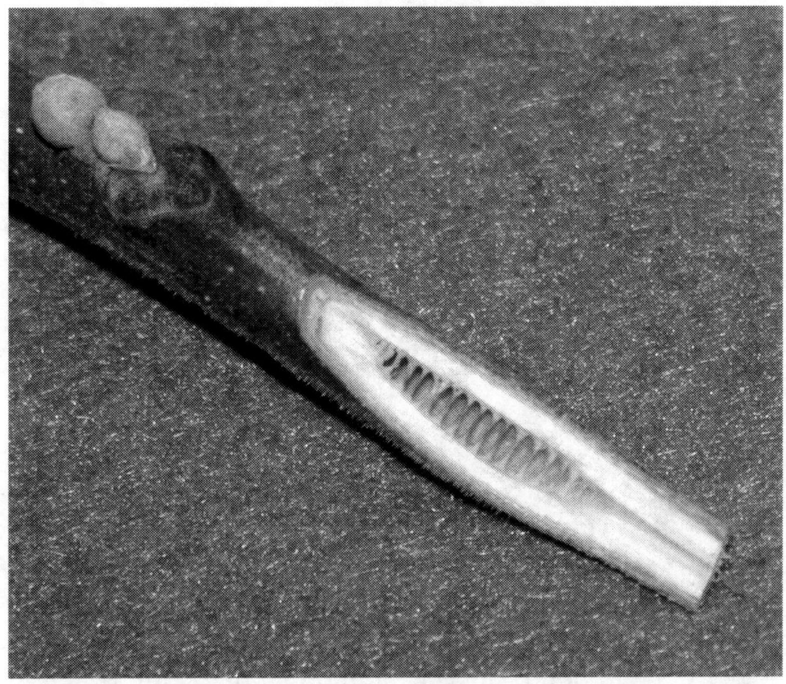

FIGURE 5.1 Walnut twig showing its chambered pith that can be used to identify the tree.

only found in other species of Juglans, such as the butternut. Besides its significance as a wood tree, the walnuts themselves are important as a food item, and the husks are used as abrasives and as a source for dyes.

Walnut sapwood is nearly white in color, while the much more important heartwood is dark brown to purple. It is very hard and heavy, having good shock resistance. Other important properties include its straight grain, easy workability, and beauty. It is used extensively in gunstocks, furniture, and cabinetwork. It is also a favorite of the hobby wood turner. Due to its high price, it is often veneered for overlay on lesser woods.

Table 5.10   Black Walnut, Green Wood Properties[*]

	Metric	English
Weight	929 kg/m^3	58 lb/ft^3
Modulus of Rupture	65.503 MPa	9.50 psi
Janka Hardness	4003.20 N	900 lbf
Shear	8.412 MPa	1.22 psi

[*]Source: U.S. Department of Agriculture, Forest Service, General technical report FPL-GTR-83, *Hardwoods of North America* (1995), by Harry A. Alden.

5.11 BOXELDER *(ACER NEGUNDO)*

Boxelder is a common North American deciduous fast-growing tree that grows to 70 ft with a diameter of 3 ft. It thrives best along rivers and ponds and is common among cottonwood and willow. Boxelder wood is light, weak, close grained, and porous. Its sapwood is white to yellow, and red streaking is common, while its heartwood is yellow to brown. Boxelder is susceptible to rotting.

Although boxelder works well with both hand tools and woodworking machinery, it is considered a second-rate wood for most purposes due to its rotting tendencies and softness. It is, however, used for general woodenware, wooden boxes and crates, and inexpensive furniture. It is also used extensively as a pulp and fuel wood.

Table 5.11   Boxelder, Green Wood Properties[*]

	Metric	English
Weight	513 kg/m^3	32 lb/ft^3

Table 5.11   Boxelder, Green Wood Properties* *(cont.)*

	Metric	English
Modulus of Rupture	35.992 MPa	5220 psi
Janka Hardness	2401.92 N	540 lbf
Shear	6.481 MPa	0.94 psi

*Source: U.S. Department of Agriculture, Forest Service, General technical report FPL-GTR-83, *Hardwoods of North America* (1995), by Harry A. Alden.

5.12 BUCKEYE (*AESCULUS SPP.*)

This genus, representing 13 species, includes the true Buckeyes along with the Horse Chestnuts. Because their wood is so similar, they are included together when used as lumber. Buckeye trees grow to 70 ft with thick branching and a domed top. They prefer rich moist soil and are often found on riverbanks and in yards as ornamentals. Their nuts are a source of starch.

Buckeye sapwood is white and gradually merges into yellowish heartwood with no definite boundary. The wood is uniformly soft, straight grained, and lightweight but is difficult to carve or machine properly. Nonetheless, Buckeye is used for crates, boxes, furniture, and caskets. When available, it is also used as pulpwood in the paper industry.

Table 5.12   Buckeye, Green Wood Properties*

	Metric	English
Weight	785 kg/m^3	49 lb/ft^3
Modulus of Rupture	33.096 MPa	4.80 psi
Janka Hardness	1289.92 N	290 lbf

Table 5.12   Buckeye, Green Wood Properties* (cont.)

	Metric	English
Shear	4.551 MPa	660 psi

*Source: U.S. Department of Agriculture, Forest Service, General technical report FPL-GTR-83, *Hardwoods of North America* (1995), by Harry A. Alden.

5.13   BUTTERNUT *(JUGLANS CINEREA)* (WHITE WALNUT)

Butternut is a hardy nut tree found from southeast Canada down to Tennessee and North Carolina. It reaches 100 ft in height with a 3-ft diameter trunk, though most trees are smaller. Butternuts are in the walnut family, and their sweet, butter-flavored nuts are used in baking and are also important for wildlife.

Butternut has a narrow band of white-colored sapwood, while its heartwood is chestnut brown with red tinges. Like walnut, it is open grained and strong, but not as dense or dark colored. It is straight grained and easily worked with hand and machine tools, taking on a lustrous, rich finish. It is used for dimensional lumber, furniture, and veneer.

Table 5.13   Butternut, Green Wood Properties*

	Metric	English
Weight	737 kg/m^3	46 lb/ft^3
Modulus of Rupture	37,233 MPa	5.40 psi
Janka Hardness	1734.72 N	390 lbf
Shear	5.240 MPa	760 psi

*Source: U.S. Department of Agriculture, Forest Service, General technical report FPL-GTR-83, *Hardwoods of North America* (1995), by Harry A. Alden.

5.14 HICKORY (*CARYA SPP.*)

Hickory trees consist of some 15 to 20 species found in North American and Asia. They can obtain heights of 140 ft with a trunk diameter of 4 ft. The nuts from many species are palatable for both humans and animals and are an important food source in the woodlands.

Hickory wood has an unmatched blend of hardness, strength, and stiffness. Its sapwood is white, while its heartwood is sharply contrasted in reddish brown. It is used in tool handles, ladder rungs, baseball bats, and wheel spokes. Hickory is sometimes used in cabinetry due to its unique pleasing grain patterns. It is a difficult wood to work with due to its properties, and it often splits and cracks during drying. Hickory is also prized for firewood for its high heat value and is also used in cooking and smokewood for its flavor.

Table 5.14   Hickory, Green Wood Properties[*]

	Metric	English
Weight	1009 kg/m^3	63 lb/ft^3
Modulus of Rupture	80.672 MPa	11.7 psi
Janka Hardness	N/A	N/A
Shear	9.446 MPa	1.37 psi

[*]Source: U.S. Department of Agriculture, Forest Service, General technical report FPL-GTR-83, *Hardwoods of North America* (1995), by Harry A. Alden.

5.15 LIVE OAK (*QUERCUS VIRGINIANA*)

Live Oak is named for its yearlong foliage and is essentially an evergreen. Among the oaks, Live Oak is more sprawling, often wider than taller in mature specimens. It attains a height of over 50 ft, with a trunk diameter of 3 ft or more. Its lower limbs branch out horizontally, and even downwards to touch the ground, before curving upward. They are found from coastal Virginia south through Florida and west to central Texas.

Live Oak wood is more dense and difficult to work with as compared to other oaks. The USS Constitution (nicknamed "Old Ironsides") made use of Live Oak's curving branches for her skeleton, and White Oak for the makeup of her 21-in thick hull. As a lumber source, Live Oak is little used today but, like other oaks, it is open grained, heavy, and hard. Its sapwood is light yellow to white, while its heartwood is darker brown.

Table 5.15   Live Oak, Green Wood Properties[*]

	Metric	English
Weight	1218 kg/m^3	76 lb/ft^3
Modulus of Rupture	82.051 MPa	11.9 psi
Janka Hardness	N/A	N/A
Shear	15.237 MPa	2.21 psi

[*]Source: U.S. Department of Agriculture, Forest Service, General technical report FPL-GTR-83, *Hardwoods of North America* (1995), by Harry A. Alden.

5.16 MAPLE *(ACER SPP.)*

There are more than 120 species of Maple known worldwide, including several commercial North American species. Maple is an extremely important material for furniture and flooring and is also prized for its flavorful sap when made in to syrup and candy.

Maple trees grow to a height of over 120 ft with a diameter of 3 ft. Their leaves are symmetrical and opposing, and their seeds are enclosed in the familiar double wing shape. All North American Maple are threatened and affected by the Asian Longhorn Beetle and, when infested, they can be locally devastated. Several fungal diseases also threaten maple. Besides their use in lumber and syrup, maples are often planted as ornamental trees for their colorful foliage each autumn.

Maple wood is very strong and dense, with two species (*A. saccharum,* the Sugar Maple, and *A. nigrum,* the Black Maple) mainly in use. The sapwood of maple is commonly white with a reddish tinge, while the heartwood is reddish brown and occasionally darker. It has a very uniform texture and is easily lathe turned. Due to its dense grain, it takes a high polish and is also easily stained. It glues with good results. While the grain of maple is often straight, at times maple has very decorative patterns known as "birds eye," "fiddleback," and "curly." These anomalies attract woodworkers for their beauty and rarity and command prices several times that of straight-grained wood. Besides being used in top-line furniture and flooring, maple is used in bowling pins, chopping blocks, and baseball bats.

Table 5.16   Maple, Green Wood Properties[*]

	Metric	English
Weight	897 kg/m^3	56 lb/ft^3
Modulus of Rupture	64.813 MPa	9.4 psi
Janka Hardness	4314.56 N	970 lbf

Table 5.16   Maple, Green Wood Properties[*] *(cont.)*

	Metric	English
Shear	10.066 MPa	1.46 psi

[*]Source: U.S. Department of Agriculture, Forest Service, General technical report FPL-GTR-83, *Hardwoods of North America* (1995), by Harry A. Alden.

5.17 PERSIMMON *(DIOSPYROS VIRGINIANA)*

Persimmon trees are slow-growing deciduous trees found in eastern North America. They obtain a height of 80 ft and a trunk diameter of 2 ft. The bark is distinguishable as resembling alligator skin, having thick, square, scaly plates. Its branches are brittle and can be easily wind damaged. The fruit of Persimmon is edible by animals and also enjoyed by people.

Persimmon sapwood is white to gray brown, while its heartwood is dark brown to black. Persimmon wood is strong, stiff, hard, and heavy. It is frequently lathe turned into bowls, cups, and pool cues. It is a favorite wood for golf club "woods" and also is used in furniture. Persimmon is in the same family as African Ebony, and when Persimmon heartwood is black in color, it may be confused with it. Persimmon, however, is much more prone to cracking and is more difficult to work with. Due to its high oil content, it does not easily glue.

Table 5.17   Persimmon, Green Wood Properties[*]

	Metric	English
Weight	1009 kg/m^3	63 lb/ft^3

Table 5.17   Persimmon, Green Wood Properties [*] *(cont.)*

	Metric	English
Modulus of Rupture	68.950 MPa	10.0 psi
Janka Hardness	5693.44 N	1280 lbf
Shear	10.135 MPa	1.47 psi

[*] Source: U.S. Department of Agriculture, Forest Service, General technical report FPL-GTR-83, *Hardwoods of North America* (1995), by Harry A. Alden.

5.18 RED ALDER (*ALNUS RUBRA*)

Although there are more than 20 species of alder, only *Alnus rubra* (Red Alder) is commercially important. Red Alder is the most common hardwood in the Pacific Northwest and the largest of the North American alders, ranging from coastal Alaska to California. They can attain heights of over 120 ft with a diameter of 30 in. They have a gray, smooth bark.

Red Alder wood is white when cut but quickly turns to yellows and browns as it ages. It has no definite boundary between its sapwood and heartwood. In warm weather, it decays quickly and must be processed soon after cutting. Although not a very strong wood, it finds uses as nonstructural lumber, plywood core stock, and pulpwood. It is also popular among hobby woodworkers due to its easy workability and ability to take a good finish. Its smoke imparts a good flavor to food and is used for cooking and smoking salmon. Native Americans also made many medicines and dyes from the Red Alder.

Table 5.18   Red Alder, Green Wood Properties[*]

	Metric	English
Weight	737 kg/m^3	46 lb/ft^3
Modulus of Rupture	44.818 MPa	6.50 psi
Janka Hardness	1,957.12 N	440 lbf
Shear	5.309 MPa	0.77 psi

[*]Source: U.S. Department of Agriculture, Forest Service, General technical report FPL-GTR-83, *Hardwoods of North America* (1995), by Harry A. Alden.

5.19 RED OAK *(QUERCUS RUBRA)*

The Red Oak is found in eastern North America from southern Canada south to Georgia and Alabama. It can grow to a height of over 120 ft with a trunk diameter of over 3 ft. The Red Oak is one of the fastest growing oaks, and it matures in as little as 20 years. It is a favorite ornamental and shade tree, displaying vibrant foliage colors in autumn. In the forest, its acorns make up a large portion of mast, feeding many animals from deer to squirrels.

The sapwood of the Red Oak is white to yellow, while its heartwood is reddish brown. Oak is coarse, open grained, strong, and heavy. It machines well and is easy to finish. Historically, oak has been used to build ships, but today it finds more domestic uses including furniture, cabinetry, and paneling. Oak barrels are used to age various wines and spirits, which gives them unique color, aroma, and flavor. Oak is also used in railroad ties, flooring, mining timbers, and pallets. It is used extensively as a fuel wood, as it burns hot and slow. A cord of oak will yield around 16 million Btu of heat, equaling 220 gal of propane.

Table 5.19 Red Oak, Green Wood Properties[*]

	Metric	English
Weight	63 lb/ft^3	1, 009 kg/m^3
Modulus of Rupture	57,229 MPa	8.30 psi
Janka Hardness	4,448.00 N	1000 lbf
Shear	8,343 MPa	1210 psi

[*]Source: U.S. Department of Agriculture, Forest Service, General technical report FPL-GTR-83, *Hardwoods of North America* (1995), by Harry A. Alden.

5.20 SWEETGUM (*LIQUIDAMBAR STYRACIFLUA*)

Sweetgum trees occur along the east coast from New York to Florida, and west to Texas. They grow to a height of 100 ft with a trunk diameter of 3 ft. Sweetgum leaves are star shaped, having five pointed lobes arranged in a nearly symmetrical pattern. They have spiked seedpods that resemble small maces as they lie beneath their long stems. Sweetgum is a major source for shikimic acid, which is refined into pharmaceuticals including Tamiflu®. Shikimic acid is found throughout the entire tree but is most abundant in its seedpods, which are harvested while they are still green. Sweetgum foliage is very ornamental, turning golden to red in autumn.

The sapwood of the Sweetgum is white to pink, while its heartwood is reddish brown. Sweetgum is a strong, hard wood having an attractive interlocking grain. It is a good material for wood turning and is workable with hand tools. On an industrial scale, sweetgum is used for veneers, plywoods, and dimensional lumber as well as for crates, pallets, and cooperages. It is a favorite deciduous pulpwood and is often planted specifically for that purpose.

Table 5.20   Sweetgum, Green Wood Properties[*]

	Metric	English
Weight	801 kg/m^3	50 lb/ft^3
Modulus of Rupture	48.955 MPa	7100 psi
Janka Hardness	2669 N	600 lbf
Shear	6.826 MPa	990 psi

[*]Source: U.S. Department of Agriculture, Forest Service, General technical report FPL-GTR-83, *Hardwoods of North America* (1995), by Harry A. Alden.

5.21 SYCAMORE *(PLATANUS OCCIDENTALIS)*

The Sycamore tree is found in the eastern U.S.A. from the Canadian border south to the gulf states. It reaches a height of over 120 ft, with a trunk diameter of 3 ft, and has very large maple-like leaves. Its bark grows slower than the tree itself and exfoliates in large irregularly shaped sheets, leaving a mottled gray-brown-white pattern. Sycamore prefers rich, moist soil and often grows along riverbanks. On larger specimens, the trunks are often hollow.

Sycamore sapwood is white to light yellow, while its heartwood varies from light to dark brown. It has a close-textured interlocking grain and is used in furniture, flooring, veneer, and plywood. Sycamore glues well and can have some very attractive fiddleback grain patterns, making it appealing for the woodworker.

Table 5.21   Sycamore, Green Wood Properties[*]

	Metric	English
Weight	833 kg/m^3	52 lb/ft^3
Modulus of Rupture	44.818 MPa	6.50 psi
Janka Hardness	2713.28 N	610 lbf
Shear	6.895 MPa	1.00 psi

[*]Source: U.S. Department of Agriculture, Forest Service, General technical report FPL-GTR-83, *Hardwoods of North America* (1995), by Harry A. Alden.

5.22 WHITE OAK (QUERCUS ALBA)

The White Oak is found in the eastern United States from southern Maine to northern Florida. It is the most widely used and important commercial oak wood and is also used as a windbreak and ornamental. It grows to over 120 ft tall, with a trunk diameter of over 3 ft, and is a slow growing, long-lived tree. Its bark is light to dark gray with shallow plates (Fig. 5.2). The acorn of the White Oak is edible and even palatable after being boiled. It is one the sweetest acorns and can be used either whole or ground up into meal.

The sapwood of the White Oak is white to light yellow, while its heartwood is darker to brown. Oak is coarse, open grained, strong, and heavy. It machines well and is easy to finish. Histori-

FIGURE 5.2 The deep furrows on the bark of the white oak.

cally, oak has been used to build ships, but today it finds more domestic uses including furniture, cabinetry, and paneling. Oak barrels are used to age various wines and spirits, which gives them unique color, aroma, and flavor. Oak is also used in railroad ties, flooring, mining timbers, and pallets. It is used extensively as a fuel wood and burns hot and slow. A cord of oak will yield around 16 million Btu of heat, equaling 220 gal of propane.

Table 5.22   White Oak, Green Wood Properties[*]

	Metric	English
Weight	993 kg/m^3	62 lb/ft^3
Modulus of Rupture	57.229 MPa	8300 psi
Janka Hardness	4714.88 N	1060 lbf
Shear	8.618 MPa	1251 psi

[*]Source: U.S. Department of Agriculture, Forest Service, General technical report FPL-GTR-83, *Hardwoods of North America* (1995), by Harry A. Alden.

5.23 YELLOW POPLAR (*LIRIODENDRON TULIPIFERA*)

Yellow Poplar trees are the tallest deciduous trees in the eastern U.S.A., with heights of over 160 ft recorded, and having a diameter of 8 ft. It is found in the eastern United States west to Illinois and, rarely, west of the Mississippi. It is fast growing and has dark-colored, deeply furrowed bark. One common name for it is the *Tulip Tree* for its multicolored cup-shaped flowers that bloom in the late spring. It is shade intolerant and prefers fertile soils.

Yellow Poplar sapwood is white and sometimes striped brown. Its heartwood is usually tan but may take on dark green hues. The wood is straight grained, lightweight, and very uniform in texture, making it popular for carving. It is one of the easiest woods to work with when using hand and power tools, and it is used in the hobby shop for furniture, toys, and general woodworking. In larger mills, Yellow Poplar is made into veneers, plywood, pallets, and shipping crates. It is also an important pulp stock.

Table 5.23   Yellow Poplar, Green Wood Properties[*]

	Metric	English
Weight	609 kg/m^3	38 lb/ft^3
Modulus of Rupture	41.370 MPa	6.00 psi
Janka Hardness	1957.12 N	440 lbf
Shear	5.447 MPa	0.79 psi

[*]Source: U.S. Department of Agriculture, Forest Service, General technical report FPL-GTR-83, *Hardwoods of North America* (1995), by Harry A. Alden.

6

Softwoods

6.1 BALD CYPRESS *(TAXODIUM DISTICHUM)*

The Bald Cypress grows in the southeastern U.S.A. from coastal Virginia south through Florida to eastern Texas. It grows mainly in swampy wetlands, often rising directly out of a continuous water cover. The Bald Cypress can reach heights of over 120 ft, with trunk diameters of over 10 ft. It is very long lived, having several specimens aged at well over 2000 years. Bald Cypress roots often reach upward through the soil and water, protruding in growths described as "knees."

The narrow band of sapwood is white, while the heartwood color varies from light yellow to darker brown and red. The wood has a combination of high strength, density, and hardness, as well as excellent decay resistance. These properties make it valuable for posts, beams, and similar timbers used in exterior and even marine applications. Uses include railroad ties, trestle timbers, fence posts, and docking. The knees are a popular base for wood carvers, who like them for their tight grain and high density.

Table 6.1   Bald Cypress, Green Wood Properties[*]

	Metric	English
Weight	817 kg/m^3	51 lb/ft^3
Modulus of Rupture	45.5 MPa	6600 psi
Janka Hardness	1730 N	390 lbf
Shear	5.58 MPa	809 psi

[*]Source: U.S. Department of Agriculture, Forest Service, General technical report FPL-GTR-83, *Softwoods of North America* (1995), by Harry A. Alden.

6.2 BALSAM FIR *(ABIES BALSAMEA)*

The Balsam Fir ranges from southeast Canada to the northeast United States. It typically reaches a height of 60 ft with a trunk diameter of 1.5 ft, although under ideal conditions it can be much larger. It can grow in elevations of up to 6000 ft. The Balsam Fir is a favorite short-needle Christmas tree, and it is an important food source for both the moose and the ruffed grouse. Its needles are among the most fragrant of the evergreens and are used for aromatic sachets.

The wood of the Balsam Fir is soft and lightweight and has good splitting resistance. Its color varies from pale yellow to light brown. Balsam Fir works easily with hand and machine tools and finishes easily. Although mainly a pulpwood, it is also used for light construction, paneling, and crates.

Table 6.2   Balsam Fir, Green Wood Properties[*]

	Metric	English
Weight	721 kg/m^3	45 lb/ft^3

Table 6.2   Balsam Fir, Green Wood Properties[*]

	Metric	English
Modulus of Rupture	37.9 MPa	5497 psi
Janka Hardness	1290 N	290 lbf
Shear	4.55 MPa	660 psi

[*]Source: U.S. Department of Agriculture, Forest Service, General technical report FPL-GTR-83, *Softwoods of North America* (1995), by Harry A. Alden.

6.3 COASTAL REDWOOD (*SEQUOIA SEMPERVIRENS*)

Coastal Redwood trees are among the largest and oldest living species on earth, with a recorded height of over 379 ft, a trunk diameter of over 25 ft, and an age of over 3000 years old. Their natural range is a thin strip from Monterey, California, north to southern Oregon, but they have been successfully transplanted worldwide, with notable stands in New Zealand and Europe. There are also successful plantations in California, where it is cultivated for its prized wood.

Its name comes from the color of its heartwood, which is light red to deep mahogany. Its wood is coarse, straight grained, and dimensionally stable, working well with both hand and power tools. Redwood, especially the heartwood, is very resistant to insects and decay. It is a relatively strong wood, and large timbers are cut for structural use in buildings, bridges, and railroad trestles. On smaller scales, it is used for its beauty in hot tubs, saunas, and decks and for interior trim work. It also splits easily, making it useful for shingles and shakes. Lesser-quality wood is used for pulp and particle board also.

Table 6.3   Coastal Redwood, Green Wood Properties[*]

	Metric	English
Weight	801 kg/m^3	50 lb/ft^3
Modulus of Rupture	51.7 MPa	7498 psi
Janka Hardness	1820 N	410 lbf
Shear	5.52 MPa	800 psi

[*]Source: U.S. Department of Agriculture, Forest Service, General technical report FPL-GTR-83, *Softwoods of North America* (1995), by Harry A. Alden.

6.4 DOUGLAS FIR *(PSEUDOTSUGA MENZIESII)*

The Douglas Fir is a western species found in spotty areas from central Mexico north to central Canada, east from the Colorado Rockies, and west to the Pacific coast. There are two subspecies: *menziesii,* the coastal variant, and *glauca,* which is found more inland. The Douglas Fir is one of the largest conifers, reaching a height of over 300 ft with a trunk diameter of over 8 ft; but these sizes are uncommon, with typical sizes usually around half this much. Its bark is thick, rough, and dark brown. The Douglas Fir is a common plantation tree, not only for its timber but also for its fast-growing, nicely shaped Christmas trees.

The sapwood of the Douglas Fir is white to light yellow, while the heartwood is yellow to light brown and red. The wood is hard, very stiff, and strong, and it is difficult to work with hand tools. It is a favorite wood for dimensional lumber and timbers used in construction. It is also popular for milled products, furniture, and even flooring, due to its hardness relative to other conifers.

Table 6.4   Douglas Fir (Coastal), Green Wood Properties[*]

	Metric	English
Weight	610 kg/m^3	38 lb/ft^3
Modulus of Rupture	53.1 MPa	7702 psi
Janka Hardness	2220 N	500 lbf
Shear	6.20 MPa	900 psi

[*]Source: U.S. Department of Agriculture, Forest Service, General technical report FPL-GTR-83, *Softwoods of North America* (1995), by Harry A. Alden.

6.5 EASTERN RED CEDAR (*JUNIPERUS VIRGINIANA*) (EASTERN RED JUNIPER) AND WESTERN JUNIPER (*JUNIPERUS OCCIDENTALIS*)

The Eastern Red Cedar is found in the eastern half of North America from southern Ontario to the gulf states. It can grow up to 120 ft tall, but it is an extremely slow-growing tree, and more often it is found at about half that height. It has a reddish brown bark, and its wood, needles, and berry-like cones are extremely aromatic. Although this tree is commonly called a cedar, it is officially known as a juniper. Despite its slow growth, it is planted as a windbreak and ornamental species.

The sapwood of the Eastern Red Cedar is thin and white, while its heartwood is deep red to brown. It is straight grained, with a fine, uniform texture that is easily worked with both hand and power tools. Due to the wood's aromatic quality, it is used in linen and clothing chests, wardrobes, and closet veneers. It is also very insect and decay resistant, making it useful for fence posts and other exterior applications. The berries of the eastern red cedar are used to flavor gin, and the distilled oils from the wood and needles are used in medicines and perfumes (Fig. 6.1).

This should not be confused with the Western Red Cedar, a western species of juniper, *Juniperus occidentalis,* which occurs in the Pacific coastal areas from California to Washington and east to Idaho. It prefers a high, mountainous elevation and is

FIGURE 6.1 The fragrant berries of *Juniperus virginiana,* or the Eastern Red Cedar.

generally not found below 2500 ft. Its structure, properties, and uses are similar to those of the Eastern Red Cedar.

Table 6.5   Red Cedar, Green Wood Properties[*]

	Metric	English
Weight	529 kg/m^3	33 lb/ft^3
Modulus of Rupture	57.9 MPa	8398 psi
Janka Hardness	2580 N	580 lbf
Shear	8.20 MPa	1189 psi

[*]Source: U.S. Department of Agriculture, Forest Service, General technical report FPL-GTR-83, *Softwoods of North America* (1995), by Harry A. Alden.

6.6 EASTERN WHITE PINE *(PINUS STROBUS)*

The Eastern White Pine is native to eastern North America from southern Canada south to Georgia and west to Wisconsin. It commonly grows to a height of over 100 ft, with diameters of over 5 ft, but occasionally can grow much larger. Several specimens have been recorded to live well beyond 400 years. Its bark is smooth and gray when young, aging to deeply furrowed rectangular plates. It is the largest eastern U.S. evergreen and was extensively used as sailing ship masts both in America and Europe. Today it is an important timber tree as well as an ornamental and reforestation species.

The sapwood of the Eastern White Pine is whitish to pale yellow, while the heartwood is light brown to reddish. It has straight grain, possesses medium strength, and has a uniform texture. It is also easily kiln dried without excessive warpage or shrinkage. Lumber from the Eastern White Pine is used in general construction, furniture, paneling, and interior finish. Its sap is used in turpentine production and, historically, as a waterproofing tar.

Table 6.6   Eastern White Pine, Green Wood Properties[*]

	Metric	English
Weight	577 kg/m^3	36 lb/ft^3
Modulus of Rupture	33.8 MPa	4902 psi
Janka Hardness	1290 N	290 lbf
Shear	4.60 MPa	667 psi

[*]Source: U.S. Department of Agriculture, Forest Service, General technical report FPL-GTR-83, *Softwoods of North America* (1995), by Harry A. Alden.

6.7 LODGEPOLE PINE *(PINUS CONTORTA)*

The Lodgepole Pine is found in western North America, from Alaska south to California and west throughout the Rocky Mountains. They typically grow to a height of 70 to 80 ft with a trunk diameter of around 16 in, but at times growing to over 50 ft with only a 3-in diameter trunk. There are four subspecies, which can be found in elevations from sea level up to over 11000 ft, growing well in all soils and conditions.

Early North American settlers, as well as native peoples, used the Lodgepole Pine for teepee frames and cabin logs. It was also a common wood for mining timbers and railroad ties. Today, it has many uses, including fence posts, utility poles, and log siding, but mainly it is part of the lumber group called spruce-pine-fir (SPF) that is used in structural framing. Lodgepole Pine sapwood and heartwood are nearly indistinguishable from each other in color, with no definite boundary. The wood is easily worked with tools, is easy to glue, and holds nails well.

Table 6.7   Lodgepole Pine, Green Wood Properties[*]

	Metric	English
Weight	625 kg/m^3	39 lb/ft^3
Modulus of Rupture	37.9 MPa	5497 psi
Janka Hardness	1470 N	330 lbf
Shear	4.69 MPa	680 psi

[*]Source: U.S. Department of Agriculture, Forest Service, General technical report FPL-GTR-83, *Softwoods of North America* (1995), by Harry A. Alden.

6.8 LONGLEAF PINE (SOUTHERN YELLOW PINE) *(PINUS PALUSTRIS)*

The Longleaf Pine spends the first 3 to 12 years of its life as a small ground cluster resembling a patch of grass. During this time, it is developing its taproot, which will extend several feet into the ground. Upon staring its external growth, it will rise 3 to 5 ft in its first year, carrying the cluster to the top of its new growth. The longleaf pine can grow to an excess of 100 ft, but this height is uncommon today, as mainly secondary growth exists. Its range is in the eastern United States from coastal Virginia south through most of Florida and west to eastern Texas.

The Longleaf Pine tree is fire resistant in both the ground cluster stage as well as in its mature form. As such, it overtakes burned-out areas better than most species can.

Once upon a time, the Longleaf Pine was a valuable source for turpentine, pitch, and timber in the shipping industry. Today, it is a popular wood for utility poles, construction lumber, and heavy timbers used in bridges, railroad ties, and beams. Longleaf pine is a very strong wood, relative to other conifers, and it is not easy to work with hand or power tools. It is not a favorite wood for interior work or carpentry.

Table 6.8　Longleaf Pine, Green Wood Properties[*]

	Metric	English
Weight	881 kg/m^3	55 lb/ft^3
Modulus of Rupture	58.6 MPa	8499 psi
Janka Hardness	2620 N	590 lbf
Shear	7.17 MPa	1040 psi

[*]Source: U.S. Department of Agriculture, Forest Service, General technical report FPL-GTR-83, *Softwoods of North America* (1995), by Harry A. Alden.

6.9 PACIFIC YEW (*TAXUS BREVIFOLIA*)

The Pacific Yew is native to the Pacific Northwest from northern California north to coastal Alaska. A second range exists in the northern Rockies from British Columbia south to Idaho, Montana, and eastern Washington and Oregon. It grows as a shrub or small tree and can reach a height of 40 ft, with a trunk diameter of 2 ft. The tree is very slow growing and has thin, scaly bark often red or purplish in color. Unlike other conifers, it has a fleshy red fruit that contains its seeds.

The wood of the Pacific Yew is extremely hard and dense to the point of equaling typical hardwoods. Its sapwood is light in color, while its heartwood is dark brown to orange. It has a very fine, straight, dense grain that polishes to a high luster. It works easily with tools and is a very popular wood for lathe turning and carving. Traditionally, and even today, it is a favorite wood for making longbows due to its strength, which can meet or even exceed that of maple.

Table 6.9   Pacific Yew, Green Wood Properties[*]

	Metric	English
Weight	865 kg/m^3	54 lb/ft^3
Modulus of Rupture	69.6 MPa	10,094 psi
Janka Hardness	5110 N	1150 lbf
Shear	11.2 MPa	1624 psi

[*]Source: U.S. Department of Agriculture, Forest Service, General technical report FPL-GTR-83, *Softwoods of North America* (1995), by Harry A. Alden.

6.10 RED PINE *(PINUS RESINOSA)*

The Red Pine is a northeastern variety, found from Manitoba east to New Brunswick and Newfoundland, and south to Minnesota through the northeast U.S.A. to Maine. It can grow to a height of 80 ft, with a trunk diameter of 3 ft. Its bark is rough and flaky, with tinges of orange to red. Through its range, it is a plantation tree grown for its timber and also for paper pulp. It is also planted for wind and snow blocks, usually along roadways or around housing.

The sapwood of the red pine is white to yellow, while its heartwood is reddish to brown. It has a straight, even grain with a medium texture, having an oily feel and resinous odor. The red pine usually grows tall and straight, making it valuable for pole and piling stock. Other uses include cabin logs and large construction timbers. Due to its deep color, it is popular as a milled wood for use in cabinetry, furniture, and trim work.

Table 6.10   Red Pine, Green Wood Properties[*]

	Metric	English
Weight	673 kg/m^3	42 lb/ft^3
Modulus of Rupture	40.0 MPa	5801 psi
Janka Hardness	1510 N	340 lbf
Shear	4.76 MPa	690 psi

[*]Source: U.S. Department of Agriculture, Forest Service, General technical report FPL-GTR-83, *Softwoods of North America* (1995), by Harry A. Alden.

6.11 SITKA SPRUCE (*PICEA SITCHENSIS*)

The Sitka Spruce is native to the western Pacific from northern California north to the coastal Alaskan peninsula. The largest of the spruce trees, it grows to over 200 ft tall with trunks of over 5 ft in diameter. Its bark is gray, scaly, and very thin—often less than 1 in thick, even on very large trees. Besides its use as a timber tree, its sap is boiled into a beverage, and its root bark has traditionally been made into woven baskets.

The sapwood of the Sitka Spruce is white to yellow, and the heartwood is a darker yellow to brown. The wood is straight grained, even textured, and very strong for its weight. It works easily with hand and power tools and kiln dries without difficulty. As a lumber wood, Sitka Spruce is used in furniture, barrels, and milled products such as trim and doors. Due to its high strength-to-weight ratio, it is popular in small aircraft and turbine blade construction. It has particular acoustic qualities that make it important for guitar and piano sound boards. One final use for the Sitka Spruce is in pulp, where its long, strong fibers and ease of processing make it valuable.

Table 6.11   Sitka Spruce, Green Wood Properties[*]

	Metric	English
Weight	529 kg/m^3	33 lb/ft^3
Modulus of Rupture	39.3 MPa	5700 psi
Janka Hardness	1560 N	350 lbf
Shear	5.24 MPa	760 psi

[*]Source: U.S. Department of Agriculture, Forest Service, General technical report FPL-GTR-83, *Softwoods of North America* (1995), by Harry A. Alden.

6.12 WESTERN HEMLOCK (*TSUGA HETEROPHYLLA*)

The Western Hemlock is found from southern Alaska south to northern California along the Pacific coast. A second population occurs farther inland from British Columbia south to northern Idaho and Montana. Like many other Pacific conifers, the Western Hemlock reaches great heights, sometimes surpassing 200 ft with a trunk diameter of over 8 ft. Its bark is thin and scaly with shallow ridges. The Western Hemlock is very shade tolerant, and when larger trees block sunlight out, young hemlock can remain nearly dormant for decades. During this time, they grow very little but extend out their branches and leaves in an umbrella shape in an attempt to capture all available light.

The heartwood and sapwood are both white in color, with a purplish tinge. Dark streaks may be present in the wood that are caused by the maggot of the *Cheilosia burkei* fly. This discoloration does not affect the wood's strength, but it detracts from its eye appeal for interior work. Hemlock is a strong, moderately light, all-purpose wood. The best grades are used for furniture and interior trim work, while the bulk of it us used in general construction and framing. Lower-grade hemlock is also used for pulp stock. The bark of the hemlock is high in tannic acid, and it was used traditionally to tan hides and produce purple dyes. It was also a favorite carving wood of the native peoples.

Table 6.12   Western Hemlock, Green Wood Properties[*]

	Metric	English
Weight	657 kg/m^3	41 lb/ft^3
Modulus of Rupture	45.5 MPa	6599 psi
Janka Hardness	1820 N	410 lbf
Shear	5.93 MPa	860 psi

[*]Source: U.S. Department of Agriculture, Forest Service, General technical report FPL-GTR-83, *Softwoods of North America* (1995), by Harry A. Alden.

6.13 WESTERN RED CEDAR (*THUJA PLICATA*)

The Western Red Cedar is a large, towering tree found along the coastal Pacific from northern California north to Alaska. It grows to over 200 ft tall, with trunk diameters of over 16 ft. The trunks of larger trees are often buttressed and fluted. It is a very long-lived tree, at times attaining ages of over 1000 years. This tree is very popular as a domestic hedgerow tree and will grow throughout most of the continental U.S.A. in varying soil and moisture conditions.

More important, however, is the wood itself. Larger specimens will contain several thousands of board feet of very high-quality lumber, known for its straight grain, pleasing color, aroma, and rot resistance. It has a very thin, white sapwood, leaving its remainder for the valuable heartwood, which is used mainly for shingles, siding, decking, and hot tubs. It is not the strongest of woods, however, and is not usually called on for heavy structural use.

Table 6.13   Western Red Cedar, Green Wood Properties[*]

	Metric	English
Weight	433 kg/m^3	27 lb/ft^3
Modulus of Rupture	35.8 MPa	5192 psi
Janka Hardness	1160 N	260 lbf
Shear	5.31 MPa	770 psi

[*]Source: U.S. Department of Agriculture, Forest Service, General technical report FPL-GTR-83, *Softwoods of North America* (1995), by Harry A. Alden.

6.14 WHITE FIR *(ABIES CONCOLOR)*

The White Fir is native to the western United States south to northern Mexico. In ideal conditions, it can reach a height of 180 ft, with a trunk diameter of 6 ft. Both the sapwood and heartwood are whitish to red-brown, and the wood has a straight, medium to coarse grain structure. The bark is whitish gray and generally smooth, and the needles have a bluish tinge on both the underside and top. The White Fir has a natural pyramid shape and is very popular in the western states as an ornamental yard tree and windbreak. It is also very popular as a Christmas tree, and is farmed extensively.

The wood of the White Fir is soft and coarse but valuable for noncritical uses including plywood, paneling, and pulp. Its lack of odor makes it useful for fruit and vegetable crates as well.

Table 6.14 White Fir, Green Wood Properties[*]

	Metric	English
Weight	753 kg/m^3	47 lb/ft^3
Modulus of Rupture	40.7 MPa	5903 psi
Janka Hardness	1510 N	340 lbf
Shear	5.24 MPa	760 psi

[*]Source: U.S. Department of Agriculture, Forest Service, General technical report FPL-GTR-83, *Softwoods of North America* (1995), by Harry A. Alden.

7

Tropical Woods

7.1 AFRICAN BLACKWOOD (*DALBERGIA MELANOXYLON*)

The African Blackwood is found in sub-Sahara Africa and grows to a height of 50 ft. It is a slender tree, rarely growing larger than 12 in in diameter. The trunk is fluted with deep impressions, is rarely round, and has papery gray to brown bark. Its roots are traditionally used in medicines.

African Blackwood heartwood is dark gray to purplish with black streaking and a sharply contrasting sapwood that is pale yellow to white. It has a fine, even texture, with straight grain and an oily feel to it. African Blackwood is a hard, heavy, and brittle wood with excellent tonal qualities that qualify it for the construction of high-end woodwind instruments. Along with the Gaboon Ebony, the African Blackwood is a commonly used dark wood for small turnings on the lathe, including bowls, chess pieces, and pipe stems.

Properties Listed at 12% Moisture[*]	
Bending Strength	31,000 psi
Modulus of Elasticity at 1,000 psi	2,980
Maximum Crushing Strength	10,800 psi

[*]Source: U.S. Department of Agriculture, Forest Service, *Tropical Timbers of the World* (1980), by Martin Chudnoff.

7.2 BRAZILIAN ROSEWOOD (*DALBERGIA NIGRA*)

True to its name, the Brazilian Rosewood's range is restricted to the single country and is not found elsewhere. It is a widely copied wood, and several inferior species exist that are being sold under the name "rosewood." The true Brazilian Rosewood tree grows to over 120 ft with a trunk diameter of 4 ft, and it is often buttressed. Although older specimens usually have a hollow trunk, the best-figured wood comes from the inside of these hollow trees.

The heartwood is colored brown to deep red, with darker striping throughout. The sapwood is very light yellow to white and very contrasting to the darker heartwood. Brazilian Rosewood is a very hard, heavy, and strong wood. When cut, it emits a very fragrant odor, giving it its name. It machines well and is also easy to cut with hand tools. True Brazilian Rosewood is one of the most costly woods due to its harvesting restrictions. The only available wood is old stock that has been cut before the restrictions have been placed, or wood that has been smuggled out of Brazil. It is a popular wood for guitars and piano cabinets. Other uses include high-end furniture, cabinetry, billiard cues, knife handles, and chess pieces.

Properties Listed at 12% Moisture[*]	
Janka Hardness	2,720 lb
Bending Strength	18,970 psi
Modulus of Elasticity at 1,000 psi	1,880
Maximum Crushing Strength	9,600 psi

[*]Source: U.S. Department of Agriculture, Forest Service, *Tropical Timbers of the World* (1980), by Martin Chudnoff.

7.3 BULLETWOOD *(MANILKARA BIDENTATA)* (BALATA)

The Bulletwood is found in Puerto Rico and the West Indies as well as Central America down to northern South America. It grows to over 150 ft in height, with trunks sometimes reaching 6 ft in diameter. Besides being a timber tree, the Bulletwood is tapped for its latex to be used in gums and rubbers. The tree also bears plum-sized edible yellow fruit.

Bulletwood heartwood is light brown to red that deepens in color when exposed, while its sapwood is yellow to light brown. It is a very hard, strong, and heavy wood that is used locally where cut for general construction, bridges, and railroad timbers. Bulletwood is also used in fine furniture and cabinetry for its pleasing colors, durability, and ease of working. Its straight grain and good strength makes it a good wood for steam bending, for use on boats. It is difficult to glue, so other methods of joining should be used.

Properties Listed at 12% Moisture[*]	
Janka Hardness	3,190 lb
Bending Strength	27,280 psi
Modulus of Elasticity @ 1000 psi	31,450
Maximum Crushing Strength	11,640 psi

[*]Source: U.S. Department of Agriculture, Forest Service, *Tropical Timbers of the World* (1980), by Martin Chudnoff.

7.4 COCOBOLO *(DALBERGIA RETUSA)*

Cocobolo is found on the Pacific side of Central America from Panama to southern Mexico. It is a medium-sized tree of poor form, and it reaches heights of 60 ft with a trunk diameter of 2 ft. Larger specimens can have hollow trunks. This species prefers dryer tropical highlands.

Cocobolo heartwood is light and variable when freshly cut but soon darkens to oranges, reds, and violets upon exposure to UV light. The wood has an oily texture and feel and has straight to interlocking grains. It machines well with both hand and power tools and finishes beautifully. It does not take glue very well, so other methods of joining must be used. Cocobolo dust is an irritant to the skin and lungs, and protection should be worn. It is used in general turnery for bowls, duck calls, pen barrels, and other small items. It is also carved into jewelry boxes and used for inlays. Cocobolo wood is a favorite material for the hobby wood turner.

7.5 EAST AFRICA OLIVEWOOD *(OLEA HOCHSTETTERI)*

Common in Kenya and Zaire, the East African Olivewood reaches a height of 120 ft with a trunk diameter of 3 ft. Trees of this size are rare, and most specimens are much smaller. Olive trees can live for periods of over 1000 years and are among the oldest trees on earth. The importance of their fruit and oil far outweigh their value as a timber tree, but their beautiful grain patterns and colors are very popular with woodworkers.

Properties Listed at 12% Moisture[*]	
Janka Hardness	1,840 lb
Bending Strength	25,300 psi
Modulus of Elasticity at 1,000 psi	2,530
Maximum Crushing Strength	12,200 psi

[*]Source: U.S. Department of Agriculture, Forest Service, *Tropical Timbers of the World* (1980), by Martin Chudnoff.

7.6 GABOON EBONY *(DIOSPYRUS CRASSIFLORA)* (AFRICAN EBONY)

The Gaboon Ebony is found in equatorial west Africa, mainly along riverbanks. It can attain a height of 60 ft with a trunk diameter of 2 ft, but most trees are harvested long before they reach this size.

Ebony heartwood varies from dark brown to black, while its sapwood is contrasting with lighter grays. Its fine texture, coupled with its high density, gives it the ability to take on an excellent finish that will polish to a bright sheen. Ebony is an expensive wood, and care must be taken to avoid substitutes that are sold as ebony, such as the African Blackwood *(Dalbergia melanoxylon)*. True ebony is in the genus *Diospyrus*, which includes the Gaboon Ebony and also the Ceylon Ebony from India *(Diospyros ebenum)*. Ebony is used in carvings, black piano keys, musical instruments, and many small decorative items. The beauty of ebony wood outweighs the difficulty in working with the material. It is tough on cutting tools and readily dulls them, due to the wood's high silica content.

Properties Listed at 12% Moisture[*]	
Bending Strength	27,400 psi
Modulus of Elasticity at 1,000 psi	2,560
Maximum Crushing Strength	13,350 psi

[*]Source: U.S. Department of Agriculture, Forest Service, *Tropical Timbers of the World* (1980), by Martin Chudnoff.

7.7 GREENHEART *(CHLOROCARDIUM RODIEI)*

The Greenheart is native to Guyana and surrounding areas. It is large tree, growing to over 120 ft with a trunk diameter of 4 ft. It is almost completely immune to decay and termites, and it is resistant to marine borers and fire.

Greenheart wood, along with its resistance to attack, is strong, heavy, and hard. It is commonly used as marine pilings worldwide and, when used untreated, will last many decades in both fresh and saltwater. Planks made of greenheart also are used in marine environments for piers and docks and are unmatched by any other wood for these purposes. The heartwood is light to dark green with gray or black streaking, becoming lighter in its sapwood. It has a fine, even texture with straight grain. It can take a nice finish and polish and is sometimes used in small specialty turned items like billiard cue ends and pen barrels.

An unusual property of this wood is its near explosive reaction to outside air when it is being cut in the mill, at times fragmenting with great force. To avoid this phenomenon, greenwood logs are chained tight while being cut. Care must also be taken in the workshop, as greenwood splinters themselves are poisonous when left under the skin, often causing infection.

Properties Listed at 12% Moisture[*]	
Bending Strength	29,860 psi
Modulus of Elasticity at 1,000 psi	4,060
Maximum Crushing Strength	17,400 psi

[*]Source: U.S. Department of Agriculture, Forest Service, *Tropical Timbers of the World* (1980), by Martin Chudnoff.

7.8 GUMBO-LIMBO (*BURSERA SIMARUBA*)

The Gumbo-Limbo is found from central Florida down to northern South America. It can attain a height of 90 ft with a trunk of 3 ft. Most specimens are about half this size in both height and diameter. The bark is thin and red; it peels off in sheets as a birch would. Its leaves are smooth, deep green, and oval shaped. The resin from the Gumbo-Limbo tree is pleasantly aromatic and used to make glue and varnish. The Gumbo-Limbo tree is extremely tolerant to high winds and is used as a windbreak.

Gumbo Limbo wood is soft and lightweight. Its sapwood is nearly identical to its heartwood, and light yellow to whitish in color. It is easy to machine, but sharp tools are needed as it is easily torn. It is used in matchsticks, boxes, crates, and construction lumber. It is also used as a filler wood in plywood.

Properties Listed at 12% Moisture[*]	
Janka Hardness	230 lb for green wood
Bending Strength	4,800 psi
Modulus of Elasticity at 1,000 psi	740
Maximum Crushing Strength	1,510 psi

[*]Source: U.S. Department of Agriculture, Forest Service, *Tropical Timbers of the World* (1980), by Martin Chudnoff.

7.9 JATOBA *(HYMENAEA COURBARIL)* (BRAZILIAN CHERRY)

The Jatoba is found from South America through Central America north to Mexico. It is a common, large tree growing to over 130 ft in height with a trunk diameter of 5 ft, with large buttresses at the base. Jatoba has an edible seedpod, and its bark and flowers are important in traditional medicines.

Jatoba wood is salmon to orange color when fresh, turning russet to deep brown when seasoned. It is often streaked with dark bands and has a golden luster. Jatoba has a coarse, interlocking grain that produces a very strong, hard wood. It machines and glues well but splits easily and requires predrilling for nailing. It is a popular material for flooring, cabinetry, and furniture as well as small turnery items.

Properties Listed at 12% Moisture[*]	
Janka Hardness	2,350–3,290 lb
Bending Strength	19,400 psi
Modulus of Elasticity at 1,000 psi	2,160
Maximum Crushing Strength	9,510 psi

[*]Source: U.S. Department of Agriculture, Forest Service, *Tropical Timbers of the World* (1980), by Martin Chudnoff.

7.10 LIGNUM VITAE (*GUAIACUM OFFICINALE*)

Lignum Vitae is found in West Indies islands through Mexico, Central America, and South America. It is a small evergreen tree with heights of 30 ft and trunk diameters of 12 to 20 in, with a smooth-textured purple/green bark. The Lignum Vitae has been overharvested to a "threatened" status and today is protected in most countries.

The Lignum Vitae is one of the world's heaviest woods, and its density is matched by its high degree of strength. Its heartwood is greenish brown to black and contrasts sharply with its light-colored sapwood. It has a strongly interlocked grain and a very oily feel due to its high resin content. Lignum Vitae is very difficult to work and does not glue readily. Due to its hardness and oil content, it is often used as bearings and bushings that are self-lubricating. Although it may seem archaic, Lignum Vitae bearings are still used today in hydroelectric plants and diesel-electric ships for shaft bearings. Other uses include mallets, ax handles, mortar and pestle sets, and courtroom gavels.

7.11 MAHOGANY *(SWIETENIA MACROPHYLLA)* (HONDURAS MAHOGANY)

The Honduras Mahogany is found from Southern Mexico down to the northern half of South America. It can grow to over 150 ft in height with a trunk diameter of 6 ft, with much thicker buttresses. Due to overharvesting, restrictions have been placed on the taking of native trees; to keep up with demand, the Honduras Mahogany has been successfully grown in plantations in both South America and Asia. It is also known as the Big-Leaf Mahogany.

Mahogany heartwood is reddish orange to pink colored when fresh cut but turns to a much darker deep red/brown color when exposed. Its sapwood is light yellow to white in color. The grain of mahogany is at times attractively curly and wavy, and always fine in texture. It works very easily with hand and machine tools and takes on a fine finish with a high luster. Mahogany is used in fine furniture, cabinetry, boat building, and interior trim. It is a favorite wood for turning bowls, pens, and similar smaller items.

Properties Listed at 12% Moisture[*]	
Bending Strength	11,590 psi
Modulus of Elasticity at 1,000 psi	1,420
Maximum Crushing Strength	6,470 psi
Janka Hardness	740 lb

[*]Source: U.S. Department of Agriculture, Forest Service, *Tropical Timbers of the World* (1980), by Martin Chudnoff.

7.12 PINK IVORY (*BERCHEMIA ZEYHERI*)

The Pink Ivory is found from South Africa north to Zimbabwe. It often grows in stands of several trees and reaches a height of 50 ft with 18-in trunks. The edible fruit is often sold in rural African villages and is an important income source.

Pink Ivory heartwood is light pink to red in color, while its sapwood is white. Upon exposure, the heartwood darkens to deep red and orange, but UV blockers can slow this process when applied to the finished wood product. Pink Ivory has mostly straight grain and a very fine texture that will polish to a high sheen. The wood is rarely on the market; when found, it commands a very high price. It is used for small wood turnings and ornamental pieces including billiard cue ends, chess pieces, and small jewelry boxes.

7.13 PURPLEHEART *(PELTOGYNE PANICULATA)*

The Purpleheart tree grows from Mexico south through Brazil and Peru and can attain a height of 170 ft with a 4-ft trunk diameter. The Purpleheart has suffered from overharvesting, and to satisfy demand, it has been successfully farmed.

Purpleheart is a very strong, durable, and heavy wood, although its color is its main attractant. Its heartwood, starting off as light brown, turns to a vibrant purple color when exposed to UV light. Over time, with further exposure, the violet color fades to a darker brown shade but still retains some of its purple. To delay this process, UV resistant coatings can be applied to the wood. Purpleheart is used in flooring, cabinetry, and furniture making. It is an abrasive wood that quickly dulls both hand and machine tools when working with it. Nonetheless, it is a favorite material of hobby woodworkers who use it in turnery, jewelry boxes, and model making.

Properties Listed at 12% Moisture[*]	
Bending Strength	19,220 psi
Modulus of Elasticity at 1,000 psi	2,270
Maximum Crushing Strength	10,320 psi
Janka Hardness	1,860–3,920

[*]Source: U.S. Department of Agriculture, Forest Service, *Tropical Timbers of the World* (1980), by Martin Chudnoff.

7.14 TAMARIND *(TAMARINDUS INDICA)*

Tamarind was originally from Africa, and was exported from there into Asia and Mexico hundreds of years ago. Today it is both farmed and grows wild in the New World from south Florida through Central America and on into Brazil. It is also found in the West Indies, Puerto Rico, and Hawaii. The Tamarind is a large tree, growing over 100 ft tall with massive trunks of over 8 ft in diameter. Its use as timber is overshadowed by its fruit, which grows in large pods and is used in various drinks, sauces, and local recipes.

Tamarind heartwood is dark red to brown, while its sapwood is light yellow to white. It is a very strong, heavy, and durable wood. Older trees have a hollow center, making wide boards rare. Where locally grown, it is very popular for its strength and hardness, and it is used for planking, general construction, and bridge timbers. It is exported for its beauty and then used for fine furniture, cabinetry, and veneer. Due to its fine texture, it is popular in turned items and takes a very fine finish.

7.15 TEAK *(TECTONA GRANDIS)*

Teak is found in Southeast Asia from Indonesia to Thailand and Burma. It can grow to over 120 ft tall with trunk diameters of 5 ft. Teak is successfully grown on plantations and has been transplanted to Central America, where it does well as a cash crop.

Teak is one of the best-known timbers worldwide. It is noted for its weight-to-strength ratio, its durability, and its resistance to weather and insects. Other notable attributes are its pleasing color and grain, along with good workability. Teak can, however, dull tools quickly at times, due to sporadic amounts of silica in the wood. Its sapwood is pale yellow, darkening to deep golden and brown in its heartwood. Teak is used for many outdoor objects including ship and yacht adornments, flooring and decking, and outdoor furniture and piers.

Properties Listed at 11% Moisture[*]	
Bending Strength	15,400 psi
Modulus of Elasticity at 1,000 psi	1,450
Maximum Crushing Strength	8,760 psi
Janka Hardness	1,000–1,155 for dry material.
Forest Products Laboratory Toughness	116 in-lb avg. for green and dry wood (5/8-in specimen)

[*]Source: U.S. Department of Agriculture, Forest Service, *Tropical Timbers of the World* (1980), by Martin Chudnoff.

Part IV

Miscellaneous Materials

8

Ceramics and Composites

8.1 ALUMINUM OXIDE (AL$_2$O$_3$) (ALUMINA)

Aluminum oxide is the most widely used engineering ceramic material due to its low cost, wide availability, and useful properties. A simple combination of aluminum and oxygen produces the ionic bond that gives it its great hardness and strength as their oppositely charged atoms are attracted together and condense. Ruby and sapphire are two types of crystalline aluminum oxide that are found in nature and colored by impurities. These gemstones can also be synthesized in the laboratory to produce top-quality gems that are chemically and optically identical to natural stones—but without the flaws that are found in nature. When grouped together, these crystalline forms of aluminum oxide are known as corundum and are lab grown for many industrial purposes alongside of their adornment uses. Corundum is nearly as hard as diamond and, when produced optically clear, makes excellent wristwatch crystals and other scratch-resistant lenses. Other uses of corundum include high-precision bearings (as in a 17-jewel watch) and in abrasives. Both natural and man-made corundum is ground and graded according to size and bonded to paper and cloth sheets as "sandpaper" abrasive sheets and vitrified into grinding and cutoff wheels.

Powdered aluminum oxide is also used in its loose form for sandblast media and also for polishing in its finer forms. Its hardness allows it to abrade and polish nearly any surface.

8.2 SILICON CARBIDE (SiC) (CARBORUNDUM)

Silicon carbide is an extremely hard compound of silicon and carbon. It is rarely found in nature as *moissanite* but more commonly is man made in electric furnaces from silica sand and carbon. It rates between aluminum oxide and cubic boron nitride in hardness (9.2 on the Mohs scale) and as such makes excellent abrasives. Unlike aluminum oxide, however, silicon carbide wheels will readily grind tungsten carbide cutting tools. As an abrasive, silicon carbide is also powdered and graded for use in grinding compounds, abrasive sheets, and cutoff wheels.

Besides being used as abrasives, silicon carbide is further used in sintered ceramic materials including brake linings, mechanical seals, ballistic vest inserts, and many other items requiring hardness, strength, high heat capacity, and wear resistance. Silicon carbide ceramics can also be found in bearing surfaces, sandblasting nozzles, and high-temperature burners.

8.3 TUNGSTEN CARBIDE

Tungsten carbide (WC) is a ceramic compound of tungsten and carbon that is joined with a nickel or cobalt matrix into an extremely hard, tough material. With hardness equal to corundum and approaching that of diamond, WC finds many uses in the metal-cutting and mining fields. Cutting tools (Fig. 8.1) prepared from WC are made from the sintering process, where the powdered carbide (a combination of powdered tungsten and carbon) is heated in a mold to the melting point of its matrix binder. Afterward, the tool is ground with a diamond wheel to its finish. In the machine shop, carbide cutting tools are used not only for cutting extremely hard metals but also for their ability to hold an edge far longer than traditional high speed steel when

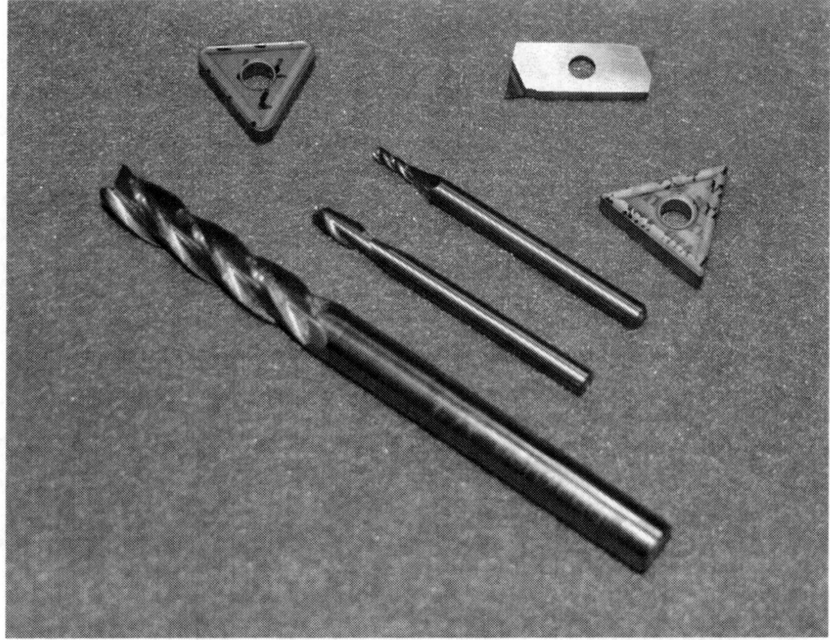

FIGURE 8.1 An assortment of solid tungsten carbide cutting tools.

cutting softer materials. Different grades of hardness are available, depending on the application. The jewelry industry has recently begun making wedding bands from WC due to the material's particular qualities. Because of its high hardness, WC jewelry will not scratch or wear easily, and it will take a very pleasing polish to a deep grayish-blue hue. Other uses include armor-piercing projectiles and bearing surfaces.

8.4 SIALON

Sialon is a family of ceramic alloys containing silicon, aluminum, oxygen, and nitrogen; its name is formed as an acronym from its ingredients. Sialon ceramics are nonporous and strong, and they have high thermal shock resistance. Like many ceramics, they can withstand high temperatures without loss of strength and provide good corrosion resistance against chemicals. Sialon's properties go beyond the typical ceramic, with its ability to handle molten nonferrous metals without them wetting or corroding. As such, they are utilized as crucibles, ladles, and other hardware used with molten metals. Their nonwetting properties allow the liquid metals to run off the tools without sticking to them or eroding them. This property is also useful in brazing, where holding fixtures made from sialon remain clean of brass runoff.

8.5 CHROMIUM CARBIDE (CR_3C_2)

Chromium carbide ceramic is an extremely hard and durable material mainly used to "toughen up" other products. One main use of chromium carbide is in the manufacture of wear plates, which are used to save the underlying material from abrasion and wear. Wear plates are made by welding a layer of chromium carbide onto a plate of mild steel. The plates are then welded (with chromium carbide welding rods) to the underlying base. Typical uses include dump trucks, mining operation conveyors, and other surfaces where rock and ore come to bear.

Besides being welded onto steel plates, chromium carbide is also sprayed with a nickel binder onto surfaces that require a combination of heat and wear resistance. The automotive and aerospace fields have many uses for chromium carbide protection, and it is used as an alternative to hard-coat anodizing on aluminum surfaces.

One final use of chromium carbide is as an added ingredient in tungsten carbide cutting tools, where it acts to retard crystal and grain growth during the sintering process.

8.6 BORON CARBIDE (B_4C)

Boron carbide has the highest hardness-to-weight ratio of any material, ceramic or other, and is the hardest material behind diamond and cubic boron nitride (CBN). Boron carbide abrasive powders are used in water-jet cutting and for general grinding and lapping processes. It can reach its melting temperature (2350°C, 4262°F) without degradation, allowing it to be cast and formed into shapes for blasting nozzles, drawing dies, and sintering dies for other ceramics. Boron carbide is also used in armor plating for vehicles and body armor for personnel where its relative light weight is a valuable attribute.

8.7 BONE CHINA

Bone china is a very fine variety of porcelain containing 25 to 50 percent bone ash. The addition of bone improves the porcelain's properties, including higher strength, whiteness, chip resistance, and translucence. Formulas vary by manufacturer, but the traditional formula is 50 percent bone ash, 25 percent kaolin clay, and 25 percent silica. Bone china was first produced in England in the late 1700s as an alternative to expensive imported Chinese porcelain. As it was found to be superior, its formula and manufacturing method spread worldwide, and it continues to be collected as the finest porcelain available. Bone china is translucent, which can be shown by holding it up to a light source.

8.8 PORCELAIN

Porcelain is a ceramic material used in tableware, vases and figurines. Clay, when fired into pottery, is known as a *refractory* (a material that retains its shape under high heat) but is very porous. Silica, on the other hand, melts at high temperatures but is nonporous and vitreous. Combining these two ingredients with added binders, fluxes, and catalysts gives the best of both materials and produces a glassy, tough article that won't sag or melt upon firing. Porcelain has been known in China for well over 1000 years and has been produced in England since the early 1700s.

8.9 CONCRETE

Concrete is a mixture of Portland cement, aggregate, and water that, when mixed, applied, and cured, is used as a construction material for buildings, roads, and other structures. It has been in use in one form or another since ancient Roman and Egyptian times, and many concrete items, including the great Roman aqueducts, still stand today.

The binder for concrete, Portland cement, is a mixture of limestone, clay, and secondary ingredients that may include bauxite, coal ash, or iron slag. These constituents are fired in a high-temperature kiln and sinter together into small nodules called *clinkers*. The clinkers are then ground into powdered Portland cement, ready to be used.

Concrete aggregate consists the gravel, stones, and sand that give it its strength. When only sand is used, that concrete is known as *mortar* and is used to adhere bricks and blocks to each other. For concrete, a mixture of sand and gravel is used. The size and shape of the gravel affect the working properties of the concrete, and also its strength, shrinkage, and durability. When gravel is not available, or when angular aggregate is needed, larger rocks are crushed and screened to size. Demolished concrete can also be crushed into aggregate for recycling, but this can be problematic as a result of contamination.

For most applications, concrete is reinforced with steel bars and mesh to compensate for its weak torsion strength. The added reinforcements act as a skeleton that keeps the concrete from cracking and ultimate failure. Reinforcing members are oftentimes put under tension before the concrete is poured around them, and the tension is released after it cures. This puts a prestress compression on the concrete that counteracts tension forces, allowing much longer and higher spanning of the concrete.

8.10 SODA-LIME GLASS

Soda lime glass is the typical glass found in windowpanes and glass bottles. It is made from a combination of silica sand, sodium carbonate (soda), and lime. Lesser ingredients are used, depending on the application and manufacturer, and include tinting and strengtheners. When the ingredients are melted in a furnace, they form into a mass that is the consistency of thick syrup, ready to be formed in one of several ways. For making windowpanes, molten glass is floated in a bed of molten tin and is spread out by gravity into evenly thick sheets. Glass may also be poured into molds for creating trinkets and statuettes similarly to methods used to cast metals. By taking molten glass "gobs" and blowing them to shape, either manually or by machine, one can form bottles and other hollow shapes. Glass can also be shaped with dies similarly to metals being forged, but taking far less pressure to do so.

8.11 BOROSILICATE GLASS

Borosilicate glass is favored for laboratory glassware due to its low coefficient of expansion, which enables it to tolerate a wide range of temperatures without cracking. It is also used as an optical glass, where it is more dimensionally stable throughout temperature variances, minimizing distortion. Borosilicate glass was at one time used under the Corning's Pyrex® name for glass bakeware, but since then, their formula has changed to a tempered soda-lime glass still sold under the same name by a different manufacturer. Corning, Inc., still manufactures several lines of Pyrex® borosilicate glasses for the laboratory, optical sciences, and other applications where a large amount of thermal expansion or contraction cannot be tolerated.

Borosilicate glass is made from standard glass recipes with around 5 to 15 percent of boron oxide included. Its coefficient of expansion, being 3.25×10^{-6} K^{-1}, is approximately one third that of soda lime plate glass. Both Corning Inc., and Schott, North America, Inc., manufacture a glass having near-zero coefficients of expansion. Zerodur®, by Schott, has an expansion of less than 0.01×10^{-6} K^{-1}, while Corning's ULE 7971 has an expansion of 0.05×10^{-6} K^{-1}. They are used for high-end telescopic mirrors and lenses.

8.12 LEADED GLASS (LEAD CRYSTAL)/ LEADED ACRYLIC

In the field of art glass, the addition of lead (between 15 and 35 percent lead oxide by weight) increases the refractive index of the glass. Unleaded soda lime glass has an index of around 1.55, while a heavy leaded crystal may have an index of 1.82 or more. With a higher index, the glass is able to break up white light into a rainbow of colors, which makes the glass object sparkle. By comparison, a diamond has a refractive index of 2.42. Hand cutting the leaded glass object into sharp facets furthers the shimmering effect and is also responsible for the high prices asked for leaded glass pieces.

Leaded glass, often containing up to 60 percent lead oxide by weight, is also used in the medical fields of radiology and nuclear medicine. As in solid bricks, highly leaded glass has adequate density to absorb X-radiation, allowing it to be used as a see-through window in the X-ray room while protecting the observer.

As a substitute for lead glass in the X-ray room, leaded acrylic polymers are also available that are lighter in weight and less prone to breakage than their glass counterparts. Leaded acrylic is also fashioned into facemasks and eye goggles for spot protection.

8.13 Amorphous Metals

In its liquid form, a metal lacks a basic crystalline structure and is as homogenous as glass. Upon hardening into a solid, however, metals will arrange their molecules into a three-dimensional lattice of crystals that define many of the metals properties. Amorphous metals, on the other hand, will retain the lack of crystalline structure from a liquid into its solid phase and remain in that condition. Also known as *glassy metals* or *metallic glass,* these new materials offer combinations of properties unavailable elsewhere. The lack of crystalline lattice in these metals give them greater strength (some surpassing a yield strength of 250 ksi) and malleability over their regular counterparts. This is accomplished by the random arrangement of the molecules that does away with atomic vacancies and voids where failure occurs. The lack of crystals also means a lack of grain boundaries, which will prevent most corrosion even when the metal is left bare. Amorphous metals also lack a eutectic melting point; they just soften and lose their viscosity as their temperature rises, as thermoplastics do. This property makes them applicable to processes such as thermoforming and blow molding, which is normally associated only with plastics.

Different alloys are being produced, and the field of amorphous metals is still in its infancy relative to other branches of metallurgy. Present applications are mostly experimental and novel, as high costs and a lack of engineering analysis on these new materials prevent their widespread use.

8.14 FIBER-REINFORCED PLASTICS

Fiber-reinforced plastics (FRPs) are composite materials containing glass, carbon, or similar fibers in a matrix of epoxy or other thermosetting resin. When glass fibers are used, the material is usually known simply as *fiberglass*. Depending on the application, the glass fibers are either loose or woven. Loose fibers can be added to the polymer matrix and the resultant poured into molds, while woven sheets and strips of glass are laid up into shape over or under the resin that will soak the glass. In either of these two methods, very large, strong structures can be built, including yachts, shelters, and housing shells. On a smaller scale, automobile panels, canoes, motorcycle helmets, and piping can all be made from glass reinforced resins.

Carbon-reinforced resins are another member of the FRP family, having advanced properties over fiberglass. Carbon fiber is much stronger and lighter than glass fiber, as are the composites made from it. Uses include high-end racing automobile panels, boats, and bicycles. The aerospace and sporting goods industries use carbon-reinforced resins for many products ranging from the Boeing 787 airframe to fishing rods.

Other fibers are used for specialty work, including Kevlar® fiber. Kevlar is similar to carbon in most aspects, but with added abrasion resistance and flexibility.

8.15 HIGH GRAVITY COMPOUNDS

High gravity compounds (HGCs) are metal/polymer composites that can be extruded and cast like traditional plastics. HGCs can be made with dense metals such as tungsten and can serve as lead substitutes for firearm projectiles, fishing sinkers, and ballasts. Less-dense materials are used where the plastic extrusion process is necessary or cost effective for the desired part, thus eliminating costly secondary machining operations. HGCs can also be cast into intricate shapes at much lower temperatures than liquid metals while still maintaining the appearance and heft of a solid metal. Depending on the metal filler powder used, densities can run from 2 to 11 g/cc, giving designers a wide berth for many products. Each specific resin used also has its own properties, including melting (softening) temperatures, viscosity, and product finish. Various resins used include nylon, polypropylene, polyethylene, and PVC.

8.16 WOOD-PLASTIC COMPOSITE

Wood-plastic composites (WPCs) are combinations of wood flour and thermoplastic resins that are mixed with lubricants and pigments and either extruded or cast into usable lumber shapes. As the plastic encapsulates the wood fibers, WPCs are rated for exterior use and often used in decking, window and door trim, park benches, and and railings. They have a usable life span from 20 years upwards. The lumber is rot resistant and straight, and it does not splinter. Its color extends through the material, eliminating the need for paint. Some disadvantages of WCPs are their initial high cost, their lower stiffness compared to that of wood, and their inability to be painted if a color change is desired. Although rated for decades of exterior use, they will slightly swell over time as well. New shapes and uses are coming into use, including construction sill plates, kitchen cabinets, and formed roofing shingles.

8.17 METAL MATRIX COMPOSITES

Metal matrix composites (MMCs) are composite materials having a matrix of a parent metal with a reinforcing material filler. As an MMC is a composite, the two materials used remain chemically and physically unchanged, giving the it the properties of both materials. The most common metal used in the matrix is alloyed aluminum, due to its low weight, high strength, and high thermal conductivity. Other metals used include titanium, magnesium, and copper, depending on the exact properties required. Silicon carbide, aluminum oxide, boron carbide, or carbon fibers are often used as the reinforcers.

The improved properties of MMCs over straight metals include their higher tensile and yield strength at both normal and elevated temperatures, higher stiffness (Young's modulus), an increase in abrasion resistance, lower weight, and increased fatigue resistance. Three types of reinforcer shapes (long fibers, short fibers, and granules) are used in the matrix. The specific components used and the ratio of matrix to reinforcement depend on the desired properties of the MMC. Various uses of MMCs include brake rotors, engine pistons, bicycle frames, and aeronautical components.

8.18 Engineered Wood Products

Engineered wood products include various man-made composites consisting of wood fibers and veneers bound by an adhesive resin. There are several types, each with specific engineered advantages over natural wood, but two of the main considerations are their improved structural properties and their ability to utilize wood biomass for construction purposes.

8.19 Laminated Veneer Lumber

Laminated veneer lumber (LVL) uses veneers of solid wood that are glued with high-grade adhesives into large structural members that have much greater strength and straightness than solid sawn wood. Moreover, they can be produced in thicknesses and spans unobtainable from a solid wood. Unlike plywood, the grain of LVL veneers runs in a single parallel direction, enabling it to carry its highest load perpendicular to its span. LVLs are commonly used in large-span doorways and other structural areas having a high load at a long length.

8.20 PLYWOOD

Plywoods, like LVLs, are made from thin veneers of wood that are laminated and glued together. They differ, though, in the fact that in plywood, the veneers are glued perpendicular to each other, giving even strength in two axes. It is available in various woods, grades, and thicknesses depending on the application. Marine grades are available that are void free and made from harder woods and waterproof adhesives. They are commonly used in boat transoms. For interior work, plywood is available with either "one good side" or "two good sides" and can be purchased with expensive hardwood veneers for cabinetry and similar work. The bulk of plywood, however, is used for residential and commercial construction work and can be found as roof and exterior sheathing, subflooring, and other structural panels.

8.21 PARTICLEBOARD

Particleboards are composite wood products made from chipped or shredded wood particles and a glue binder that is formed, compressed, and trimmed into sheets. Particleboard is generally for interior use, as it is not moisture resistant to any degree. Wood veneers or melamine laminates are often applied to the exterior of particleboard to increase aesthetics, protect against moisture, and make cleaning easier. Its main use is in domestic and commercial furniture, including desks, bookcases, display cases, and television stands. Other uses include shelving, countertops, and kitchen cabinets.

8.22 ORIENTED STRAND BOARD

Oriented strand board (OSB), like particleboard, is made from wood flakes and a resin binder. It differs, however, in that OSB contains generally larger strands of wood that are laid in an oriented, one-direction pattern. This gives the boards added strength in one direction over unoriented particleboard. Another difference is that OSB wood strands are "encapsulated" in wax and the resin binder, which improves its moisture resistance considerably. The main function of OSB is as a plywood substitute for construction sheathing, where it is used as roof, wall, and subfloor applications at a lower cost than plywood.

8.23 HARDBOARD

Hardboard is a high-density engineered fiber board manufactured under great heat and pressure. Very small fibers, with no grain alignment, are used in hardboard, resulting in a smooth, homogeneous, flat surface that is favored by artists, furniture builders, handymen, and architects alike. It makes a great surface for painting and takes glue and nails well. Masonite brand hardboard was a popular exterior house siding during the 1980s and into the 1990s, and it was well liked for its low cost, ease of application, and good looks. Over time, however, the siding deteriorated due to weather, and it was found that Masonite was not suitable for sustained exterior use. The siding was susceptible to rotting, buckling, and swelling and was replaced at the expense of the Masonite Co.

Today, a large market exists for hardboard in the automotive field, where it is used in auto headliners, visors, sub-interior door panels, and spare tire covers.

9

Adhesives

9.1 ANAEROBIC ADHESIVES (LOCTITE®)

Anaerobic adhesives are a group of bonding agents that cure in the absence of oxygen with the help of metal ionization. As such, they are used mainly as thread-locking adhesives that can replace lock washers and prevent loosening through vibration and shock. Anaerobics are available with varying degrees of strength, depending on fastener size and the possible need to remove the fastener at a later date. Special formulas are used that will fill gaps up to 0.010 in (0.254 mm) on cylindrical fits and can be used to "fix" a loose pin or enlarged hole. When used in manufacturing, this type of anaerobic can do away with tight tolerances on such fits, reducing costs.

9.2 BONDO® PUTTY

Bondo is a two part synthetic system that, when mixed into a putty, will cure and harden over time, enabling it to be used in a wide variety of applications—most specifically, as an auto body filler. After hardening, Bondo can be sanded and painted to blend in with the original panel. The resin used in Bondo body filler is a compound of polyester resin, styrene, and talc, with several lesser ingredients. The hardener is composed mainly of benzoyl peroxide, water, and benzoic acid. When mixed, cross-linking of the molecules occurs, and the thermoset mixture begins its one-way path to solidity. As with other thermosetting resins, Bondo polyester putty cannot be melted and reformed after curing.

Aside from its primary use as an auto body filler, Bondo is also used for mending dents and defects in other products, including furniture, machine tools, and marine vessels. It readily adheres to wood, most metals, fiberglass, stone, and concrete without shrinking, and it is generally waterproof. It is also used around the house for patching rotten spots on door and window jams.

9.3 CAULK

The term *caulk* refers to a variety of materials used to seal joints and gaps in different materials and structures after assembly. In the building trades, caulk is used to seal doors, windows, and other openings from wind and weather, and also as waterproofing in sinks and baths. Caulks of this type are made from silicone, latex, polyurethane, or similar materials and cure from a liquid to a rubberlike consistency, allowing them to be applied accurately from a tube where needed.

Another type of caulk is *oakum,* which is used in wooden ships to seal the space between planks and timbers. It is made from natural fibers, including hemp and jute, that have been infused with pine tar, creosote, or asphalt. Oakum is pressed into place with a knife or chisel and topped with tar. As the ship will slightly buckle and sway while at sea, the resilience of the oakum keeps the joints tight.

Oakum is also used to seal the joints of cast iron pipes. It is wedged into place and then covered with liquid lead. Once the lead hardens, it is caulked into place with a special iron and hammer, effectively sealing the joint.

9.4 CONTACT CEMENT/RUBBER CEMENT

These two similar adhesives are applied by a brush, roller, or sprayer to both surfaces that are to be joined, then allowed to dry before assembly. When assembled, contact cement will be permanent, and further alignment and positioning will be impossible. Rubber cement, on the other hand, is more lenient in both regards. Items fastened with it can be repositioned or completely dismantled, and the glue gummed away.

The difference in the two lies in the nature of the suspended adhesive. Rubber cement uses natural rubber latex and is available with either water or solvent carriers. Contact cement, on the other hand, is made from stronger synthetic rubbers, mainly neoprene, and may also contain phenolic resins. Like rubber cements, it is available in both water and solvent bases, including toluene, acetone, and various hexanes. Water-based contact and rubber cements, although not as strong as their solvent-based counterparts, are a good alternative when working in enclosed areas where fresh air is unavailable, or when an open ignition source may exist, as the solvents are heavier than air and may creep along the floor.

Rubber cement is popular with crafters and scrapbookers for positioning a variety of articles, including photographs, ribbon, and similar lightweight objects. Contact cement is often used in laying Formica® on counter tops and also for placing veneer on furniture. It will also permanently bond leather, metal, and many plastics.

9.5 CYANOACRYLATE (SUPER GLUE®, KRAZY GLUE®)

Cyanoacrylate adhesives are extremely fast setting and can bond most nonporous materials, as well as some porous surfaces including pottery and ceramic. They are available in several formulas, including some having high-temperature resistance (stable up to 500°F, 260°C), variable viscosities, added rubber strengtheners, and flexible curing times. They are great for temporarily holding various tooling, jigs, and fixtures in the machine shop; after use, the tooling can be knocked free and the glue easily scraped or burned away from smooth steel and aluminum surfaces.

Cyanoacrylates readily bond skin and, although it can be a nuisance when working with the glue, it is also a blessing in many surgical instances where the lack of traditional suture scars is desirable or when a rapid joining is needed in the field to prevent blood loss or infection. Similarly, a small smear of cyanoacrylate can be used to replace a bandage on small, superficial cuts.

When working with cyanoacrylates, their fumes readily react to the moisture left when handling objects to be glued, leaving a raised, visible fingerprint on smooth surfaces. Forensic workers, who use fumigated cyanoacrylates to more easily find and secure fingerprints for criminal investigations, exploit this reaction.

9.6 GUM ARABIC

Gum arabic is a multi-use water-soluble gum found as the hardened sap from various species of the Acacia tree, most importantly *Acacia senegal* and *Acacia seyal*. It has extraordinary solubility and binding properties not found in other substances, either natural or man-made. In its most simple use, it is found on the back of postage stamps as edible glue. More importantly, though, are its uses in the food and beverage industries, as well as in lithography and other printing arts.

As a food and beverage additive, gum arabic binds and emulsifies various ingredients into suspension and keeps them in suspension, preventing separation and layering. As a food additive, it is not only edible but also low in calories and a source of natural fiber.

In the lithography process, water-based gum arabic is spread over the original drawing to keep oil-based copy ink from adhering anywhere except directly over the original drawing lines, effectively copying the original image. The gum arabic acts as an ink repellent in all other areas of the surface.

Gum arabic is also an ingredient in water-based inks and paints, acting as a binder between the pigment and the water, keeping the components in suspension. Artists also add pure gum arabic to existing watercolor or acrylic paints to alter the properties of the paints.

9.7 HIDE GLUE

Hide glue is a formerly important adhesive for furniture, cabinetry, and general woodworking, made from collagen that has been extracted from bovine and equine hides. Once the collagen is extracted, it is dried and sold as granules, powders, and flakes. For use, it is heated and melted with added water, and kept hot during use. As the glue cools, the water evaporates and binds the wood together. Hide glue is not water resistant, but if the furniture it holds is kept dry, it can keep indefinitely—or at least for several hundred years, as witnessed by U.S. colonial and earlier European furniture.

Furniture joints adhered with hide glue have the advantage of easy repair over modern glue joints. When a hide-glued wooden member requires replacement, it is only a matter of heating the joint; the glue will melt, allowing the wood member to be removed. When replacing it, the added hot hide glue will melt and combine with any of the old remaining glue, making a joint as good as new. Modern polyvinyl acetate white and yellow glues will not do this, and all dried glue of this type must be removed from the joint being repaired and reglued.

9.8 HOOK AND LOOP (VELCRO®) FASTENERS

Mimicking the hooked burr of the burdock plant, Velcro uses two different surfaces that readily adhere to each other. One of the two surfaces uses an array of small nylon or similar material hooks that latch into small loops on the other surface. Velcro has been used to replace buttons, zippers, and many other mechanical fasteners in products ranging from shoes and tents to machine accessories and automobile headliners. It has the advantage over traditional fasteners in its ease and speed of use, requiring only light pressure to adhere the two sides together.

Swiss engineer George de Mestral, who noticed the effects of the cocklebur on his clothing, invented Velcro in the 1940s. In 1955, Switzerland granted de Mestral a patent for his invention, and his hook and loop fastening system gained worldwide fame.

9.9 HOT MELT ADHESIVES (HOT MELT GLUES)

Hot melt glues are available in stick form for use in handheld glue guns, or in pellet, flake, and block form for industrial systems. The adhesives cure by simply cooling from a liquid to a solid and require no solvent evaporation or mixing. Hot melt adhesives are available in a variety of thermoplastic materials intended for either general or material-specific use, and they vary by melting temperature, setting time, viscosity, and tack. Other properties include long-term flexibility, stability, and chemical resistance. Specific blends are formulated with added plasticizers, UV blockers, and tackifiers.

Hot melt can be used as a substitute for many different types of glues. It has important advantages, including fast setup time, which can reduce floor space and man hours of labor in the assembly department. Hot melt also lacks VOC pollutants, which eliminates poor air quality and related regulatory requirements.

Handheld glue guns are popular in the home for craft, hobby, and repair use. The fast set time makes clamping unnecessary but still gives sufficient time to position and reposition an item, unlike cyanoacrylate (super) glues.

In industry, hot melt glue's many uses include bookbinding, tennis shoes, cardboard boxes, and automobile and aircraft trim and moldings. For high-volume work, hot melt flows through heated tubes to either a handheld delivery device or a stationary nozzle, where the work moves underneath.

9.10 POLYVINYL ACETATE (WHITE GLUE, ELMER'S® GLUE)

Polyvinyl acetate, or PVA, is the familiar white glue used in home and school. Its many favorable properties, including nontoxicity, washability, low cost, and ease of application make it popular. White glues do not cure but simply lose their moisture through evaporation upon exposure to the air; i.e., they dry. They can be partly reconstituted with water after drying, making the joints formed by these glues not at all waterproof. PVA glues are popular with bookbinders, crafters, and hobbyists, and they are generally used on paper, foam, and wood products.

Chemically similar are the yellow carpenter's glues that develop a faster tack than their white counterparts. Although these glues are PVA based and will clean up with water, their proprietary formulas are guaranteed to be waterproof upon drying. Wood joints made from both yellow and white PVA glues are stronger than the wood itself. Glued joints can be simply clamped together and require no nails, screws, or other fasteners. Excess glue can be planed or sanded away when dry, leaving a clean, smart joint.

10

Carbon Forms and Fuels

10.1 GRAPHITE

Graphite is a soft, black, hexagonal form of pure carbon that is both found in nature and produced synthetically. It is electrically and somewhat thermally conductive, black in color, and opaque to light. Graphite has a greasy feeling to the fingers due to its lubricating properties, and it smudges easily. One final asset of graphite is its refractory properties, enabling it to withstand temperatures of up to 3000°F (1649°C) for extended periods, and up to 5000°F (2760°C) in the case of arc electrodes.

Natural graphite, found in flake or nearly amorphous forms, is an important material for raising the carbon content of steel in the foundries. Carbon, when introduced to steel, raises its hardenability, enabling it to perform tasks it could not do in its soft state. Another major use of natural graphite is in lubricants, either dry or in a grease binder. Dry graphite powders are used to lubricate mechanisms that would be otherwise contaminated by grease or oil that could collect and harbor debris. As an additive to petroleum grease, graphite adds superior lubricity as well as high temperature and pressure tolerances. The high refractory properties of graphite make it useful for crucible containers for molten metals and glass and for laboratory work. Graphite, mixed with clay, gives us the "lead" in pencils. Different formulations, of course, give different hardnesses and shades.

Synthetic graphite can be made more chemically pure and consistent than can be found in nature. A major use of this graphite is in electrodes for steel electric arc furnaces, which are mainly used to recycle steel. In its operation, a high-voltage arc is initiated from the electrodes and passes to the steel underneath. The high temperature of the arc melts the steel, readying it to be cast into new products. Graphite is the only readily available material that is electrically conductive and has the high heat capacity needed for the electrodes. Synthetic graphite offers tighter tolerances in factors such as electrical resistance, which would alter the arc if the resistance varied. Other uses of synthetic graphite include fuel cell plates, carbon fiber plastics, and radar absorbent materials.

10.2 ANTHRACITE

Anthracite is the hardest, densest, and most "energy-packed" type of coal. With a carbon content of over 90 percent and a capacity of 15,000 Btu per pound, it is the most valuable and expensive type of coal. Its appearance is glossy and reflective, having an almost metallic luster. Its main use is in home heating, but other uses include water and sewer filtering media and the production of charcoal briquettes for outdoor cooking. Anthracite, after further refining, is also a source of carbon for the manufacture of synthetic graphite and is used in both ferrous and nonferrous metallurgy.

The state of Pennsylvania, the by far highest U.S. producer of anthracite, had an output of 1.7 million short tons in 2009, according to the U.S. Energy Information Administration (EIA). Also, according to the EIA, Pennsylvania has a demonstrated reserve base of anthracite at of over seven billion short tons, giving these coal beds a projected lifespan of over 4000 years at current production levels.

10.3 BITUMINOUS

Bituminous is the workhorse of the coals, being the most commonly mined and burned type. It has a carbon content ranging from 50 to 88 percent and an energy capacity of 10,000 to 15,000 Btu per pound (Fig. 10.1). Bituminous, like other coals,

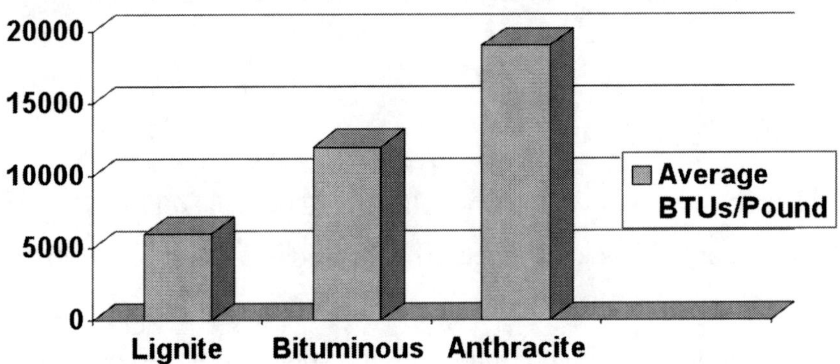

FIGURE 10.1 Energy densities of coal.

has a vegetative origin. As plant life decays and is buried under sand and soil, it accumulates and is compressed over time. The production of coal follows a natural transformation, over hundreds of millions of years, from plant life, to peat, to lignite, to bituminous, and finally to anthracite, providing that the necessary heats and pressures have been present.

A main use of bituminous coal is in the production of coke, where powdered coal is heated in an oven that is absent of oxygen. This process drives off the impurities, leaving a nearly pure form of carbon. At the elevated temperatures in the oven, the coal liquefies and, upon solidification, becomes very light and porous, having a specific gravity of only 0.7 or so. It then exits the oven and is air or water quenched as its final form: coke.

Bituminous coal is also a common fuel for electricity production. Simply stated, the coal is burned to produce heat that transforms water into steam. The steam is then used to power turbines, which are geared directly with electric generators. Nearly half (44.5 percent) of the 3.9 trillion kW·h of electricity produced in the United States in 2009 was made from coal, according to the U.S. Energy Information Administration.

10.4 Lignite

Lignite is a soft, brown to black coal with a relatively low energy content as compared to other coals, often at 5000 to 8000 Btu per pound. It has an average carbon content of 35 to 45 percent. It is in an intermediate stage between peat and harder coals, often having plant and wood remnants visible in its makeup. Its low energy content gives it limited uses, and transporting it long distances is often too expensive to be viable. As a source of home heating fuel, it is dirty, smoky, and rarely a practical option. In its range, however, it is used as a fuel for electricity production. As it is often found nearer to the surface than other coals, it can be extracted at lower costs, producing inexpensive electricity in its proximity. The Lignite Energy Council states that 79 percent of all lignite mined in the U.S.A. goes to electricity production, with the balance going to the production of synthetic natural gas and fertilizer products. Synthetic natural gas, colloquially known as *syngas* or *producer gas,* is produced by a reaction of incandescent lignite to steam and oxygen under high pressures. This technology is centuries old but has been refined over the years to produce cleaner gasses at higher efficiencies than old time "coal gas" that was once used to light city streets. Gasification can also be accomplished with charcoal produced from hardwood, giving an alternative fuel source for the automobile.

10.5 PEAT

Peat is the remains of decayed vegetation that has formed in a low-oxygen environment, often used as a source of fuel for home heating and occasionally for the production of electricity. Peat is found in marshes, swamps, and bogs and usually has a moisture content of over 50 percent. In preparation for use, it is cut into manageable-size slices either by hand or by machine and air dried in the sun. It is considered a renewable energy source and is popular in areas where wood and coal are absent. Peat is also further refined into briquettes as an alternative to cut slabs. The briquettes are very low in moisture and burn with a clean, smokeless flame. It is the smoke from peat, however, that gives Scotch whiskey its distinctive flavor, as the malted barley is dried over natural peat fires.

Besides being a source of fuel, peat has other uses. When dried, it is very moisture absorbent, making it useful as a plant-growing medium and as a soil improver. Dried peat also has the capacity to absorb its own volume in oils and fuels, enabling it to be used as a spill absorber either in loose form or in sock and tubes. It is also used as a bedding for farm animals.

Peat bogs are often drained to facilitate extraction. At times, these dried bogs ignite by accident, lightning strike, or even spontaneous combustion, producing smoldering fires that even burn underground and are impossible to extinguish. One such occurrence plagued Russia in 2010, where more than 30 peat fires joined several hundred forest fires in producing unprecedented deadly smog throughout the western half of the country.

10.6 Wood Pellets

Made from both hardwood and softwood sawmill and wood-lot waste, wood pellets have become popular as a fuel source to heat homes and businesses. Burning wood pellets has a comparable cost to burning raw, seasoned oak, with easier handling, less tendering, and less mess. Industry standard wood pellets have a density of 40 lb/ft^3 and a moisture content of 4 to 8 percent.

In the production of wood pellets, the wood scrap is dried, hammermilled, and extruded into pellet form. The extrusion process produces sufficient heat and pressure for the wood particles to adhere together with no added binders needed. Wood pellets are sold retail in 40-lb bags or delivered in loose form to customer-owned silos.

Wood pellet stoves are equipped with hoppers to load the pellets and augers to feed them into the burning chamber. They require electricity to run the auger and heat circulating fans and, due to the fact that pellet stoves do not radiate heat like traditional wood stoves, they are ineffective in a power outage unless a supplemental form of electricity is available.

In addition to wood pellets, corn pellets or even dried corn can be burned in wood pellet stoves. Corn tends to leave more ash and clinkers than wood pellets do, but they provide an alternative fuel in times of low wood pellet supply.

10.7 COKE

Coke is a purified form of carbon fuel usually extracted from coal. In its production, powdered coal is heated in an oxygen-free oven to a temperature of around 2000°F (1200°C). At this temperature, the volatile compounds of the coal are driven off, leaving mostly pure carbon behind. The gaseous by-product compounds, consisting of various syngasses and chemicals, are collectively known as *coal gas* and are piped away for other uses, either in house or after being sold to other industries. The remaining carbon at this point is in a spongy, near-liquid mass. As it leaves the oven, it is either air or water quenched to a cool temperature. Most coke production is done on sight in steel manufacturing plants, and here the completed coke is moved to the smelters.

Coke, as a final product, is used to melt and reduce iron ore for the production of steel (Fig. 10.2). If the iron ore is of good quality, it can be simply broken up into small (1 to 1.5-in) pieces and sent directly into the blast furnace with the coke. Limestone is also added to act as a flux as it forms a slag that collects impurities. A hot blast of air ignites the coke, which in

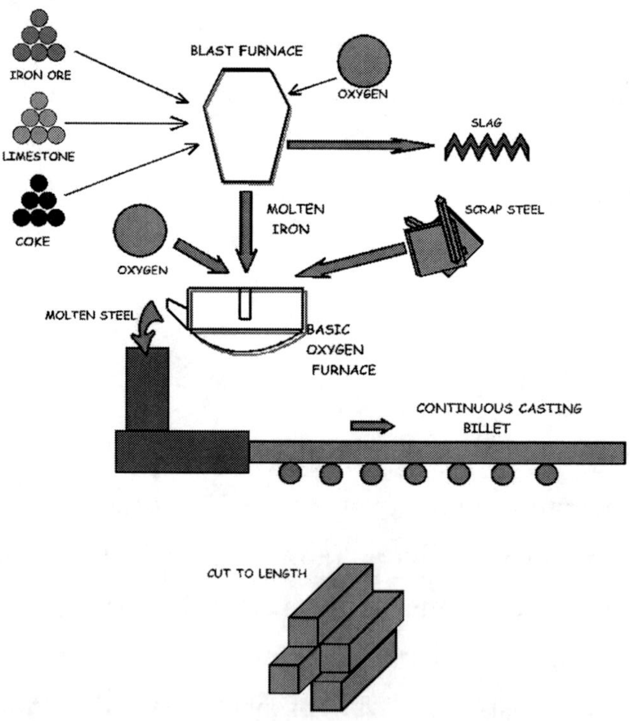

FIGURE 10.2 A simplified diagram of steelmaking.

turn burns hot enough to extract the iron from the ore. The blast furnace is a continuous process; the molten iron is periodically drawn off from underneath while the raw materials are fed from the top. Volatile gasses are also drawn off, scrubbed, and used as a fuel to heat the incoming air. The liquid iron that exits the furnace is either solidified into pig iron or immediately carried off to be further refined into steel.

10.8 CHARCOAL

Charcoal (Fig. 10.3) is the resultant carbon material left over from the burning of wood in an oxygen-free atmosphere, in a process called *pyrolysis*. In this manner, most of the water and volatile compounds are driven off, leaving a very clean solid

FIGURE 10.3 Charcoal.

fuel that burns hot and smoke free and is used mainly for outdoor cooking. Charcoal is produced in two distinct ways: either as natural block (commonly called *lump* charcoal) or as briquettes. Lump charcoal, once cooked, is not further processed and is ready for sale and use. It lights easily and is by-product free. Briquettes, on the other hand, are made from wood scrap and sawdust, along with powdered coal and various binders and ignition aids that are compressed into shape and baked. They have the advantage of being a more consistent product, batch after batch, than lump charcoal, much in the same way that a blended whiskey or orange juice has, while at the same time providing new life for old sawdust.

Other uses of charcoal include filtration and purification of both liquids and gasses. For this application, charcoal is "activated" in a process that makes it extremely porous, giving it far more surface area. This allows it to either collect impurities or react to them chemically. Activated charcoal is used in chemical masks, in fish tank filters, and in purifying drinking water. It is also taken by mouth to absorb certain orally ingested poisons.

10.9 Coal Gas

Coal gas is produced from the destructive distillation of coal. It is used as a fuel source for lighting, cooking, and heating as well as a source of various constituents including pure hydrogen. In the production of coke, coal gas is produced as a byproduct and used as a source of heat in coke's own production, as well as that of steel. Coal gas, as it is extracted, contains various gasses, including carbon monoxide, carbon dioxide, methane, and hydrogen. As a source of fuel in the early Industrial Revolution, coal gas was burned in this raw form and varied by the type and purity of the coal used. Today, coal gas has been largely replaced by natural gas and propane for home, commercial, and industrial heating and cooking, but recently it has been drawing attention to itself due to the existing vast untapped coal deposits.

A process called *underground coal gasification (UCG)* can be utilized to draw coal gas from otherwise unminable deposits. In the process of UCG, an underground reservoir of coal is burned with a controlled oxygen and steam mixture, and the resulting coal gas is piped to the surface. With this fairly simple technology, the energy from coal veins can be tapped in areas where conventional mining is impossible or impractical, such as areas underneath lake and sea beds. Coal gas as used today is more refined than in days of yore, when it was laden with sulfur, nitrogen, and other contaminants. Today, it represents a small but growing market as a clean fuel for the production of electricity.

Coal can also be converted into a synthetic liquid fuel that replaces various petroleum distillates, including gasoline and fuel oil. A method for the liquefaction of coal known as the Fischer–Tropsch process (named after the co-inventors) was invented in 1920s Germany and was a major fuel source for heavy military vehicles as well as automobiles during WWII.

10.10 Oil Shale

According to the U.S. Bureau of Land Management (BLM), there are an estimated 1.23 trillion barrels of oil locked up in oil shale deposits, primarily in Colorado, Utah, and Wyoming. These deposits consist of sedimentary rocks rich in *kerogen,* an organic compound that, when extracted, can be used as a synthetic petroleum. To extract the kerogen, the shale is crushed and heated, which drives it off as a vapor. It is then condensed into shale oil that is ready to be burned as is, or it is further refined for specific purposes. When oil shale is of sufficient quality, it may be crushed and burned without further refinement, but the rock itself remains behind and must be disposed of.

Oil shale refinement into liquid fuel, although simple in principle, is an expensive process and requires very high crude oil prices to be competitive. A barrel of shale oil may take nearly half of its energy to produce it, including mining, transportation, and the required heat for kerogen extraction. Oil companies are often reluctant to invest heavily on the process, as a possible dip in crude oil prices may render their product unsalable at market prices. This happened on May 2, 1982, when Exxon cancelled a $5 billion dollar oil shale project in Colorado due to falling oil prices and let go more than 2000 workers.

10.11 Crude Oil (Petroleum)

Crude oil, or petroleum, is a flammable liquid obtained from underground reservoirs and used as a raw material for various fuels and plastics. Made up of varying ratios of carbon and hydrogen along with several lesser ingredients, crude oil is the remnant of plants and animals and has been compressed by ancient seabeds over the span of millions of years. As these seabeds lacked sufficient oxygen to completely decompose the biomass, the resultant carbon and hydrogen were left behind, becoming the building blocks for today's petroleum. As the ancient biomass varied from area to area, so now does the petroleum. Some of it is found as thick, black oil, high in sulfur and other contaminants, while others are thinner, lighter in color, and "sweeter."

Petroleum drilling occurs over land and underneath present sea floors. It is then sent by either a pipeline or a tanker ship to the refinery, where it is broken down into its useful products in a fractional distiller. Different components of petroleum will boil off at different temperatures, which is where the fractional distiller comes in. It consists of a high vertical tower with outlets at varying intervals. As the petroleum is heated inside of the distiller, the lighter products (including gasoline), which have a lower boiling temperature, will travel the entire height and exit near the top of the column. The heavier products (including asphalt) have a higher boiling temperature and are less volatile, so they exit the tower at its lower levels. Intermediate of these are the kerosenes, diesels, and fuel oils.

10.12 KEROSENE

Kerosene, also known as *paraffin oil,* is a thin petroleum distillate primarily used for heating, lighting, cooking, and jet fuel. At room temperature and all the way to up over 100°F, it is below its flashpoint and nonvolatile, making it safe to store and use relative to gasoline. Kerosene can be used as a fuel for conventional internal combustion engines, however, when preheated above its flash temperature before entering the combustion chamber. As an engine fuel, though, kerosene is mostly used in both civilian and military jet fuel, where combustion does not depend on a low flash point. Kerosene is also known as *number 1 fuel oil*.

Kerosene heaters have been in use since the 1800s and are still popular today for homes as well as temporary-use buildings like garages and workshops. The heaters are portable, have self-contained fuel tanks, and have been proven over time to be safe to use. Before electricity, kerosene was used extensively as a source of light, and today kerosene lamps are indispensable to the Amish population and in areas absent of electric power. Savvy homeowners keep lamps and fuel handy for power outages that occur during storms and rolling blackouts. By the same token, small kerosene cookers are available that can both cook a hot meal and warm a small room.

Several lesser uses for kerosene include solvents, cutting oils, and pesticide carriers. It was at one time used to treat wounds and lice and even taken internally for various conditions.

10.13 PETRODIESEL

Thicker and heavier than kerosene or gasoline, petrodiesel, or simply *diesel,* is a petroleum distillate extracted from crude oil. Named after the diesel engine and its German inventor, Rudolph

Diesel, the fuel that bears his name is used to power heavy trucks, busses, ocean-going vessels, locomotive engines, and military tanks. The diesel engine is different from conventional internal combustion gasoline engines in its ignition. While an electrical spark ignites gasoline in its engine type, diesel engines compress the fuel and air mixture to the point of autoignition by pure compression alone. These engines also run with a lot fewer parts and do not require spark plugs, a spark distribution system, or carburetors. (They do, however, require glow plugs to ignite the fuel during a cold start.) Diesel engines also develop more horsepower at lower rpm, which provides more torque at lower speeds and a wider power band. In automobiles, diesel engines are sometimes used as an alternative to gasoline engines. Diesel engines tend to have longer lives and offer better fuel economy than comparable gasoline models but, due to their higher compression ratios, diesel requires a more robust, heavier engine. They are also noisy and tend to vibrate at idle speeds.

There are different grades of petrodiesel, including a relatively new *ultra low sulfur diesel (ULSD)* blend that has a maximum of 15 ppm sulfur. New fuel standards reduce engine emissions at the cost of increased pump prices and slightly lower energy content than previous grades.

10.14 FUEL OIL

The lower petroleum distillates, when burned to produce heat, are known as *fuel oils.* Number 2 fuel oil, the most common type used in North America, is similar to petrodiesel, and the two may exit the fractional distiller at the same time. Petrodiesel, however, is further refined to extract sulfur and other contaminants that may be left behind in fuel oil. Fuel oil is also dyed red in color to signify that highway taxes have not been paid on it, supposedly discouraging truck drivers from using it as a motor fuel. In practice, however, due to lower demand, number two fuel oil has a retail price that is very similar to petrodiesel fuel, even though petrodiesel is further refined and carries a tax of over $0.50 per gallon in the United States.

10.15 GASOLINE

Gasoline, or *petrol,* is a liquid fuel distilled from crude oil. Most gasoline produced is used to fuel internal combustion gasoline engines, with a lesser percentage used as a solvent for oil-based products or as a fuel for lanterns and torches. Gasoline, as it comes from a fractional distiller, is in a rather crude form as far as the needs of a modern internal combustion engines go. Inside of the combustion chamber in an engine, raw gasoline tends to burn sporadically "ahead of itself," leading to what is known as *knocking, pinging,* or *detonation.* Various anti-knock additives are used to control combustion and prevent this phenomenon, including tetraethyl lead, methyl tertiary butyl ether (MTBE), and iso-octane. Other additives include corrosion inhibitors and lubricants in varying formulas. Ethanol, or ethyl alcohol, is often added to gasoline as an extender at levels from 5 to 15 percent. It may lower the price of the gasoline but does so at the cost of lower available energy per gallon purchased.

10.16 Asphalt (Bitumen)

Asphalt is a thick, black petroleum product produced as the "bottom of the pot" in the fractional distiller, having the highest boiling point of any petroleum distillate. It is also found in its natural form in veins of sedimentary rock, in liquid pools, and locked up in tar sands. Its makeup consists of approximately 80 percent carbon and 15 percent hydrogen, the balance consisting of minor ingredients including sulfur, nitrogen, and oxygen.

Naturally occurring asphalt has been used since antiquity as an adhesive and as waterproofing material for crafting canoes, oceangoing ships, and aqueducts. Asphalt's waterproofing qualities are used today in the construction of roofing shingles and as a coating for fence posts, but by far its most common use is in asphalt concrete.

Asphalt concrete consists of the adhesive (asphalt) with a gravel and sand aggregate that is used for road, parking lot, and runway surfacing. When compared to conventional concrete, asphalt is less expensive and faster to lay, but it has a shorter lifespan. In northern climates where seasonal buckling can occur due to frost heaves, asphalt is more forgiving than concrete and can lay back down when the heaves melt.

10.17 Natural Gas

Natural gas is a common gaseous fuel used for residential, industrial, and commercial heating and cooking as well as in electricity production and transportation. It occurs in underground pools, either on its own or sitting atop reservoirs of petroleum. Natural gas consists mainly of methane (CH_4), with several lesser gasses and contaminants that are usually present. Before natural gas is piped to its final destination, it is refined down to nearly pure methane. The usable by-products, which may include butane, propane, and ethane, are collected at the refinery and used elsewhere. The odorant *mercaptan* is added to the final product to enable the detection of leaks of this otherwise odorless, colorless gas.

Natural gas is a popular fuel in close-proximity communities where it is piped underground to individual homes and businesses. For rural areas, natural gas can also be liquefied and transported via trucks to home- or business-owned tanks. It is also liquefied for transportation from the refinery to the market, where it is regasified before piping.

Because of its lower emissions as compared to burning gasoline, natural gas vehicles are used in urban settings for busses and other public transportation.

10.18 PROPANE (LIQUEFIED PETROLEUM GAS)

Propane (a.k.a. LPG) is a gaseous fuel that is condensed to a liquid form for ease in transportation and storage. Unlike natural gas, propane is trucked to its final destination and transferred to end-use storage tanks rather than being piped underground. Propane as a fuel is used similarly to natural gas for home space heating, cooking, and water heating. In industry, propane fuels forklift trucks in factories and warehouses, and many municipalities use propane-powered fleets for city business. Farming operations rely on propane to run standby generators, to heat barns and sheds, and to dry crops like tobacco, grain, and fruit.

Propane comes from different sources, including distillation of petroleum and as a by-product of natural gas refining. As a petroleum distillate, propane is driven off at low temperatures relative to most other products. Once it is distilled, it must be separated from similarly volatile gasses, including butane, before it can be condensed for transportation and use. Odorants are added to aid in leak detecting, as in natural gas.

10.19 BIODIESEL

Biodiesel is an additive or replacement for petrodiesel made from animal or vegetable fats. As these fats are farmed and grown, the fuel is considered renewable, in contrast to nonrenewable petroleum fuels. Through a process called *transesterification,* the raw oils are reacted with methanol, in the presence of a catalyst, to produce methyl esters (biodiesel), with glycerin as a by-product. Biodiesel, as a petrodiesel replacement, has similar energy and octane values but will gel faster in low-temperature operations. Biodiesel can deteriorate natural and butyl rubber seals and hoses, but it has better lubricity properties than the new low-sulfur petrodiesels. Most of the time, instead of running straight biodiesel, it is combined in varying amounts with petrodiesel for motor fuels in a give-and-take blend.

The interest in and use of biodiesel, along with other petroleum replacements like ethanol and shale oil, rises and falls in direct relation to world oil prices. In the aftermath of the oil embargo of the 1970s, alternative fuel activity was at a high that tumbled once oil prices returned to normal levels. Again, as oil and petroleum costs go up, there is new interest and attention given to these substitutes.

10.20 Ethanol (Ethyl Alcohol) (Grain Alcohol)

Ethanol is the common spirit found in alcoholic beverages including beer, wine, and distilled liquors made from the distilled resultant of the reaction of yeasts to sugars. Besides being consumed as a drink, it is also used as a replacement fuel or additive to gasoline in internal combustion engines. As an additive, it is commonly blended at 10 to 15 percent ethanol to gasoline, at which concentration it can be burned in unconverted engines. At higher ethanol-to-gasoline concentrations, however, the fuel system must be converted to run a richer mix due to ethanol's lower energy content (only two thirds that of gasoline).

Like biodiesel, ethanol is a renewable biofuel that can be grown and produced from seasonal crops. Much debate surrounds the use of food crops for the production of ethanol, as it can drive food prices higher due to lower availability. There is also debate concerning the net energy gain of producing ethanol, considering the costs and energy involved in both crop growth and ethanol distilling. There are many crops that contain sufficient sugars or starch to be metabolized into ethanol, including sugar beets and cane, corn, potatoes, and sorghum. To start the fermentation process, yeast is introduced to the stock in an oxygen-free atmosphere, where it converts the glucose into ethanol and carbon dioxide. After fermentation, much of the water remains in the alcohol mixture and must be removed through distillation. As alcohol boils at a much lower temperature than water (173°F, or 78°C), its vapors can be drawn off in the distiller, leaving the water behind—provided that the temperature of the mash does not exceed the boiling point of water. The vapors are then condensed back into liquid ethanol.

10.21 HYDROGEN (AS A FUEL)

Hydrogen (atomic number 1, symbol H) is a lightweight, flammable, gaseous element rarely found free in nature. It is usually compounded with oxygen in water or with carbon in natural gas. Unlike conventional sources of fuel, however, hydrogen is produced *energy negative.* In other words, it takes more energy to produce hydrogen gas than can be gained when the resultant hydrogen is burned. This unbalance labels hydrogen as an energy *carrier* rather than an energy *source.* Once extracted, though, hydrogen makes an attractive fuel for automobiles having either internal combustion engines or fuel cells. When burned, hydrogen simply recombines with oxygen, producing water as a by-product.

The technology of hydrogen-fueled internal combustion automobiles has already been created and appears to be on hold pending an inexpensive source of hydrogen. To power an internal combustion engine, hydrogen must be stored compressed on board, and a method of metering the hydrogen intake is needed, much like a natural gas setup. Another use of hydrogen as a fuel is in a fuel cell, where hydrogen reacts chemically to oxygen to produce electricity that in turn is used to run an electric motor, powering the automobile. This arrangement, much like the production of hydrogen itself, has a negative energy output-to-input ratio, but fuel cells enable electricity to be produced virtually maintenance free, with no moving parts. It also enables hydrogen to carry energy from one source to another, albeit with a substantial loss.

Glossary

alloy: The resultant of two or more metallic elements combined to produce a new and separate material. Alloys often contain properties not found in the combining metals or will have certain properties more enhanced over the original elements.

amorphous: Glass-like in nature, without any discernible crystalline structure.

aqua regia: A combination of nitric and hydrochloric acid having the ability to dissolve gold. A typical mixture will contain one part nitric to three parts hydrochloric.

austenitic: A nonmagnetic iron-bearing metal, being nonmagnetic either through makeup (as in austenitic stainless steels) or through heat (as a straight carbon steel heated to the point of being nonmagnetic).

birdseye maple: Small circular figure found embedded in the grain of certain maple wood, giving it high esteem among woodworkers and end users. Its great value is brought on both by its beauty and its rarity.

blight: A disease of the plant kingdom characterized by lesions, discoloration, and wasting away that often ends in the death of the host. Blight can be caused by fungus, insects, bacteria, or other pathogens.

board foot: A unit of wood volume equaling a 12-in wide board that is 12 in long and 1 in thick, or any combination yielding 144 in^3.

brazing: Brazing is a process similar to soldering wherein a relatively low-melting-temperature filler metal is used to fuse two metals having a higher melting temperature. It differs from soldering, though, by using a filler metal (usually brass) having a much higher temperature than that of solder.

British thermal unit: A unit of energy that equals 0.293 Wh, or the energy needed to raise the temperature of 1 lb of water 1°F. Usually abbreviated Btu.

Btu: See "British thermal unit."

carburizing: A technique used to harden the exterior shell of low carbon steel by diffusing an exterior source of carbon into the metal's skin using high heat. Also known as "case hardening."

case hardening: See "carburizing."

casting: A process of pouring molten metal, plastic, or other material into a mold that is in the desired shape of the final product. Once the casting material has hardened, the mold is either separated or broken away, leaving the casting intact.

cold working: An increase in strength and hardness of a metal through plastic (nonreversible) deformation carried out at a much lower temperature than the metal's melting point.

While cold working does increase both hardness and strength, the material's ductility and malleability decrease considerably. Cold working is usually accomplished by bending, rolling, or otherwise mechanically straining the material.

compression strength: The amount of crush that a material can withstand before failure, measured as weight per square unit applied.

cement clinkers: Small nodules resulting from the firing of limestone, clay, and lesser ingredients that make up the majority of Portland cement.

cross linking: The nonreversible bonding of one polymer chain to another, usually found in thermoset plastics. Cross linking occurs in three dimensions, leaving large, strong molecules with properties that cannot be matched with traditional thermoplastics.

curly maple: See "Birdseye maple."

deciduous: Deciduous trees are those that lose their leaves annually during the autumn season, in contrast to "evergreens." Most of the North American hardwood trees are deciduous, with the most notable exception being the Live Oak *(Quercus virginiana)*.

density: A measurement of relative compactness, given as weight per cubic unit. Typical units of density include pounds per cubic inch and grams per cubic centimeter.

dimensional lumber: Lumber that is cut to specific sizes and ready for the retail market. Typical dimensional lumber sizes include common one-by-fours, two-by-fours, and four-by-fours. The dimensions given are preplaned sizes, however, and actual measurements will be less, meaning that a North American two-by-four will actually be 1.5 × 3.5 in.

ductility: The ability of a material to be stretched or drawn into thinner sections.

elastic deformation: A temporary, reversible bend or twist applied to a material that will spring back to its original shape once the force has been lifted. Coil, leaf, and torsion springs exhibit elastic deformation.

elastomer: A rubberlike substance having the ability to return to its original shape after extreme deformation, including stretching up to seven times its original length.

electrical conductivity: The ability of a material to conduct electricity through it, defined as conductance across a length. In metric units, this is expressed as millisiemens (MS) per centimeter. In English units, electrical conductivity of metals is often expressed as a percentage of International Annealed Copper Standard (IACS) or the conductivity found in commercially pure copper. Electrical conductivity is the reciprocal of electrical resistance.

element (metallic):   A pure metal with no simpler constituent components. Compare with "alloy."

elongation yield: The amount that a material can be stretched before breakage, expressed as a percentage of the original length. For example, if a 10-in length of a steel bar can be stretched to 12 in before breakage, that particular steel bar has an elongation yield of 120 percent.

eutectic: A quality of a material, usually a metal, having the same melting and solidifying temperature. While some metals go through a mossy or pasty range when going from a liquid to a solid, a eutectic metal will flash harden when its temperature is reached.

evergreen: A tree that keeps its leaves or needles throughout the winter season, in contrast to a deciduous tree, which loses its leaves annually during the autumn season. Evergreens do in fact shed and grow new leaves and needles, but they do it throughout the year in small amounts.

extrusion: A process of squeezing a pliable material through a die, either by pushing or pulling. An extruded product will have the same cross section as the extrusion die. This is a very inexpensive method for producing plastic and metal parts very quickly. Some extrusion operations are carried out with molten materials, some with heated and softened materials, and some are accomplished cold.

fatigue resistance: The ability of a material to resist repeated loading and unloading without failing. Polypropylene plastics have such a high fatigue resistance that thin sections are used as hinges (see "living hinge") and can withstand several hundred thousand cycles, if not more.

ferritic (stainless steels):   Ferritic stainless steels are magnetic, low carbon, and high chromium. They make up the 400 series stainless steels and are easily welded.

ferromagnetic: A characteristic of certain materials (mainly iron, steel, nickel, and cobalt) that display a high level of magnetic permeability, or ease of magnetic attraction.

fiddleback maple: See "Birdseye maple."

forging: The technique of forcing (hammering) a metal piece into an open-ended die. Parts produced through forging are stronger than cast, machined, or extruded pieces, but it is an expensive process and is generally reserved for when the extra strength gained is required.

formability: The ability of a material to be permanently deformed through cold working without failing. Ductile and malleable materials have good formability, while brittle ones have very little, if any.

free cutting: Any metal that, when machined, produces short, brittle chips that will not clog up the cutting tool. They require less horsepower per unit of metal removed. Free cutting metals also save operator time in not having to constantly remove long, stringy chips from the work area in a production environment. Also known as "free machining."

fusible alloys: Any low-melting-temperature metallic alloy, some having melting temperatures that are less than their parent metals.

gumminess: A nontechnical term used to describe the characteristics of soft, sticky metals when they are being machined. Gummy metals have very low machinability, tend to stick to the cutting tool, and are notoriously difficult to leave with a good finish.

hardenability: The potential for a metal to be hardened. In carbon steels, the hardenability increases directly with carbon content as well as the speed in which the metal is cooled during the quenching phase of heat treating. Parts having large cross sections, therefore, have less hardenability than those with thinner sections.

hardness: The resistance of a material to deform when a force is applied to it. For the purposes of measuring and testing the hardness of metals and plastics, indentation methods are used. Various indentation methods, including Rockwell, Vickers, and Brinell, measure the depth that a hardened stylus makes in a material when a given load is applied. The stylus (indenter) may be made of diamond, tungsten carbide, or hardened steel, depending on the relative hardness of the material tested, and also upon the method used. For use on wood, the Jenka hardness test, by contrast, measures the force needed to penetrate a 0.444-in (11.278-mm) hardened steel ball to a depth of half of its diameter.

heartwood: The dense, usually darker-colored central wood of a tree. Heartwood is often referred to as *dead wood,* as its cells do not carry water and nutrients to the upper branches. Its strength, density, and color make it substantially more important than its neighboring sapwood for use in furniture and cabinetry, especially in hardwood. The deep reds and blacks often found in tropical species come mainly from heartwood.

heat treating: The altering of a metal's properties through heat. There are several aspects to heat treating, including hardening, tempering, and annealing, each technique being used to impart a specific quality to the metal. The same metal, before and after heat treating, can differ by in these properties by several hundred percentage points, enabling one metal to function in many applications.

high speed steel (HSS):   A family of tool steels able to withstand higher heats and abrasions as compared to high carbon steels. HSS is commonly used in twist drills, end mills, and lathe cutting tools. Various HSSs are made with combinations of tungsten, chrome, vanadium, and cobalt alloyed with high carbon steel.

hot dipped galvanization: The process of dipping steel or aluminum, usually in sheet form, into a molten bath of zinc for the purpose of corrosion prevention.

hydrogen embrittlement: The phenomena of some steels, nickel, and other metals to becoming brittle upon contact with hydrogen, which can lead to cracking and total failure.

knee (cypress):   The upward growing root extension of cypress trees. As the cypress tree is often found in flooded swamplands, the knees often protrude through the water to a height of a couple of feet and give the swamps in which they grow a familiar-looking landscape.

living hinge: A thin section of plastic that serves as a hinge between one or more moving parts, wherein the hinge is molded together with the hinging elements. In choosing a living hinge, a plastic with a great degree of fatigue resistance is needed, and polypropylene or polyethylene are usually the first choices.

machinability: The relative ease in which a particular metal or particular hardness of that metal can be machined. Different factors of machinability are taken into account, including the metal's chip quality (short chips that break apart as the metal is being cut is much more desirable than long, stringy chips that interfere with the machining operations), and the power needed to remove a quantity of metal. A metal can also be either too soft and gummy, or too hard to be properly machined. Two benchmarks are often noted for their machinability. When comparing ferrous metals, alloy 1212 steel is rated at 100 percent, while, when comparing copper alloys, alloy 360 brass is given the same rating. While alloy 1212 steel has been universally accepted as the ferrous benchmark, there are in fact steels with much greater machinability, including the leaded steel alloy 12L14, having a relative rating of 170 percent that of 1212.

machining: A manufacturing technique in which a part is made by removing metal from either raw stock or a semi-finished item using machine tools including lathes, milling machines, drill presses, or grinders. Machining is the opposite of a technique such as casting, where metal is added to a mold to produce the desired item.

magnetic permeability: The relative capacity of a material to absorb magnetic energy. Materials having a high degree of magnetic permeability are used as magnetic shields in computer hard drives, electron tubes, and many other sensitive equipment and components.

malleability: The ability of a material to be hammered or rolled into thin sections without cracking. Gold is the most malleable material, either found in nature or alloyed in the lab, having the ability to be hammered into a leaf with a thickness of five millionths (0.0000005) of an inch thick.

martensitic (stainless steel): A family of stainless steels having a high chrome and carbon content, no nickel, and the ability to be heat treated.

mast: The mixture of tree nuts, berries, and other hard or soft fruits that make up forest food for wild animals.

mercaptan (methyl mercaptan):   A synthesized odorant applied to flammable and explosive gasses that would otherwise be odorless. Mercaptan is also found in nature emanating from decaying animal matter.

modulus of elasticity: See Young's modulus.

Moh's scale: A scale of hardness defined by ten common minerals, as follows: 1. talc, 2. gypsum, 3. calcium, 4. fluorite, 5. apatite, 6. orthoclase, 7. quartz, 8. topaz, 9. corundum, 10. diamond. A material, when defined by a number on the Moh's scale, will scratch the mineral below its rating, while being susceptible to being scratched by the mineral above it. For example, a steel file will scratch orthoclase and be scratched by quartz, giving the file a Moh's number of 6.5.

mortar: A mixture used to fill gaps and bind brick, stone, and block. It is generally a combination of Portland cement and sand, being similar to concrete without the larger aggregates included.

ore: A mineral-rich rock from which metallic elements are extracted. Most metals are not found in their native (pure as found in nature) form and must be extracted from various ores.

Pilkington process: A process, named after its inventor, in which molten glass is floated on a bed of molten metal, usually tin, for the purpose of producing flat, smooth panes. As the molten glass is poured onto the tin, it spreads out into an even thickness and begins to cool. It is then removed from the bath using rollers and taken to tempering furnaces where it can undergo gradual cooling.

plastic deformation: The permanent bending or breaking of a material by the application of outside forces. A crashed automobile panel is an example of plastic deformation. Compare to "elastic deformation."

plasticizers: Various ingredients added to a material to increase its suppleness or *plasticity,* either as a whole or when the material is subjected to harsh (such as excessively hot or cold) environments.

poly hinge: See "living hinge."

polymer: A material consisting of molecules strung together in long, repeating chains. The single molecules themselves are known as monomers (*mono* meaning *one*). The term *polymer* is most often applied to plastics. While both terms are often used interchangeably, there are other polymers not at all related to plastics, including proteins and starches.

pulp stock: Any raw timber product that is used for the production of pulp for papermaking. Many tree species are grown specifically for pulp, while in other trees, the less desirable specimens or the tree tops are used.

pyrolysis: A process of incineration in the absence of oxygen that decomposes organic material, liberating various gaseous compounds and leaving a concentrated carbon form. Charcoal is produced by pyrolysis.

refractive index: The ratio of the speed of light traveling in a vacuum to the speed of light traveling through the material in question. In relation to optics, a material having a higher refractive index can bend light (such as when used in eyeglasses) more concisely than a material having a lower index, which means that polycarbonate lenses can be made thinner than ones made of acrylic, due to their respective indexes.

refractory: A material that maintains its properties at high temperatures and is used to contain molten metals, glass, and similar items.

RTV: Abbreviation for *room temperature vulcanization*. Refers to a family of silicone rubbers that cure at room temperature. Type 1 RTV is a single part, self-curing rubber, whereas type 2 RTV is a two-part silicone that will cure when mixed.

sapwood: The outer wood of a tree, usually lighter in color and weaker than the inner heartwood.

shape memory: The characteristic of certain metal alloys (mainly nitinol) to remember their previous, permanent set shape when deformed and then heated. Some plastics also exhibit shape memory properties.

sintering: The process of heating a ceramic or metal powder in a mold until the grains of powder fuse together, but are not yet at the melting temperature. Like traditional cast products, sintering builds up the product in the mold, in opposition to machining, where material is removed from a larger blank.

slag: The waste by-products that float atop of steel during its manufacture. Although it may contain a small amount of elemental iron, further extraction is usually not cost effective.

soldering: The joining to two metals by heating them to the melting point of the filler metal, usually a lead alloy or one of its substitutes.

spangle: The crystalline grain structure found on hot dipped galvanized sheet. The size of the spangle crystals can vary in size, and at times be almost invisible, depending upon the exact alloy used as a coating as well as the cooling rates once the steel is dipped and removed.

spinning: The forming of a thin section of metal into cup- or bowl-like shapes by spinning the metal on a lathe and forcing it around and over a hub in the shape of the product.

stress corrosion cracking: The rapid cracking and failure of a metal component due to reactions with certain chemicals, often when the metal is under mechanical stress at high temperatures. It differs from common corrosion in its rapid onset and failure, and also in its cracking (often microscopic at first) rather than a wasting away of the metal as found in traditional corrosion.

super elasticity: The property of certain metal alloys, mainly nickel/titanium, to exhibit exaggerated elasticity over common metals, enabling them to be greatly deformed while being able to return to their original position without plastic deformation.

superalloys: A term for different families of alloys (mainly nickel based) that retain good hardness, strength, and corrosion resistance at high temperatures. Superalloys find uses in jet engine components, nuclear reactors, and other critical applications.

tempering: The drawing back of hardness by reheating a metal after the initial quenching. Using high carbon steel as an example, the part to be tempered is first heated to its austenitic phase and then quenched in oil, water, or air, where it rapidly cools and transforms into a rock-hard, brittle phase known as "martensitic." The steel is then reheated to between 300 to 600°F (≈150 to 315°C), which relieves much of the brittleness of the steel—at the cost of some hardness. When a piece of polished steel is brought to tempering temperatures, different colors are produced by the heat applied, and these colors can be used to judge the temperature that the steel has achieved.

tensile strength: The strength of a material when the material is being pulled apart, measured in pressure per square unit at failure. Common units used to measure tensile strength are pounds per square inch (psi), and megapascal (Mpa).

thermal conductivity: The property of a material to allow the passage of heat through it, measured by the amount of heat being transferred a specific distance through the material.

thermoplastic: A type of plastic that can be repeatedly heated to melting and cooled to solidifying. Compare to "thermoset."

thermoset: A type of plastic that, after solidifying, cannot be reheated to melting. When cured thermoset plastics are heated to above their critical temperature, they degrade and burn.

toughness: The property of a material to absorb plastic deformation without fracturing. It can also be described as a combination of strength and ductility, as the lack of either of these qualities will leave a metal without toughness, either being too weak or brittle. Toughness, as a property of a material, is not generally, but can be, quantified by measuring the amount of kinetic energy a material can be subjected to at failure.

transesterification: The reaction of raw bio-oils with methanol to produce methyl-esters or biodiesel.

veneer (wood): A thin layer of usually highly attractive wood that is used to cover a less desirable material.

vitrification (grinding wheels): The fusing of loose abrasive materials, such as aluminum oxide powder, into a glasslike solid by high heat and pressure.

volatile organic compound (VOC): VOCs are high pressure (easily vaporizing) carbon based chemicals that are often hazardous and can be either man-made or natural.

welding: The joining of two pieces of metals by bringing the joint temperature up to the melting point of the particular metal, at which point they fuse together. A filler metal is sometimes used to assist in the process.

Young's modulus: A measurement of elasticity or stiffness in a material and its resistance to deformation. Soft materials like silicone rubber and various thermoplastics have a much lower Young's modulus than a harder, stiffer material like hardened tool steel. It can be measured by the change in size (either compression or stretch) that a material shows once a certain pressure or force is applied.

Index

About the Author

This is the first book by author Brian Dereu, who spends his time away from the keyboard running the small family business, the "Dereu & Sons Mfg. Co.," located in Eldridge, Missouri. Brian has been in the metalworking field his entire career, and in 1996, he was able to quit his day job and successfully enter self-employment with no more than a Bridgeport milling machine and a South Bend lathe. He is the author of several magazine articles, and his switch from manufacturing to writing is still a work in progress. As his three working sons grow in the business, he finds more available time for his new passion and is not looking back.